MOLECULAR DIAGNOSTIC PCR HANDBOOK

MOLECULAR DIAGNOSTIC PCR HANDBOOK

By

GERRIT J. VILJOEN

International Atomic Energy Agency,
Vienna, Austria

LOUIS H. NEL

University of Pretoria,
Republic of South Africa

and

JOHN R. CROWTHER

International Atomic Energy Agency,
Vienna, Austria

IAEA

FAO FIAT PANIS

Springer

A C.I.P. Catalogue record for this book is available from the Library of Congress.

ISBN-10 1-4020-3403-2 (HB)
ISBN-10 1-4020-3404-0 (e-book)
ISBN-13 978-1-4020-3403-9 (HB)
ISBN-13 978-1-4020-3404-6 (e-book)

Published by Springer,
P.O. Box 17, 3300 AA Dordrecht, The Netherlands.

www.springeronline.com

Printed on acid-free paper

Printed in the Netherlands.

TABLE OF CONTENTS

PREFACE

The Joint FAO/IAEA Division of Nuclear Techniques in Food and Agriculture is involved in agricultural research and development and assists Member States of FAO and IAEA in improving strategies to ensure food security through the use of nuclear techniques and related biotechnologies, where such techniques have a valuable and often unique role. In particular, molecular diagnostic methods have rapidly evolved in the past twenty years, since the advent of the Polymerase Chain Reaction (PCR). They are used in a wide range of agricultural areas such as, improving soil and water management; producing better crop varieties; diagnosing plant and animal diseases; controlling insect pests and improving food quality and safety.

The uses of nucleic acid-directed methods have increased significantly in the past five years and have made important contributions to disease control country programmes for improving national and international trade. These developments include the more routine use of PCR as a diagnostic tool in veterinary diagnostic laboratories. However, there are many problems associated with the transfer and particularly, the application of this technology. These include lack of consideration of: the establishment of quality-assured procedures, the required set-up of the laboratory and the proper training of staff. This can lead to a situation where results are not assured.

This book gives a comprehensive account of the practical aspects of PCR and strong consideration is given to ensure its optimal use in a laboratory environment. This includes the setting-up of a PCR laboratory; Good Laboratory Practice and standardised of PCR protocols. Examples of Standard Operating Procedures as used in individual specialist laboratories and an outline of training materials necessary for PCR technology transfer are presented. The difficulties, advantages and disadvantages in PCR applications are explained and placed in context with other test systems.

Emphasis is placed on the use of PCR for detection of pathogens, with a particular focus on diagnosticians and scientists from the developing world. It is hoped that this book will enable readers from various disciplines and levels of expertise to better judge the merits of PCR and to increase their skills and knowledge in order to assist in a more logical, efficient and assured use of this technology.

James D. Dargie
Director, Joint FAO/IAEA Division
of Nuclear Techniques in Food and Agriculture
Vienna, Austria

ACKNOWLEDGEMENTS

The authors would like to thank the following people and Institutions for their contributions, suggestions and permissions:

Tom Barrett and the Institute of Animal Health, Pirbright Laboratory, United Kingdom; Marco Romito, David Wallace, Pravesh Kara, Anel Espach, Josephine Mitchel, Zilliang Wang, Amanda Dobora, Celia Abolnik, Janet Mans, Steven Cornelius and the Biotechnology Division, Onderstepoort Veterinary Institute, South Africa; Sandor Belak and the National Veterinary Institute and the Swedish University of Agricultural Sciences, Uppsala Sweden; Roche Diagnostics GmbH; PE Biosystems; Perkin Elmer and Applied Biosystems and to all those who peer reviewed this book.

Chapter 1

CHAPTER ONE
BACKGROUND

1. Aims of this book
2. What is PCR?
3. What is the use of PCR?
4. PCR and infectious diseases - the veterinary picture

1. Aims of this book

The Joint FAO/IAEA Division of Nuclear Techniques in Food and Agriculture is involved in research and development in the fields of agriculture with particular responsibility to Member States. Nuclear techniques play an increasingly valuable and often unique role in agricultural research and development. They are used for a wide range of purposes such as improving soil and water management, producing better crop varieties, diagnosing animal diseases, controlling insect pests and improving food quality and safety. As such the Polymerase Chain Reaction (PCR), with its massive potential and already proven value, is at the forefront of debate and interest to a wide range of scientists in developing countries.

The transfer of PCR[1] technology, although challenging, is often made without adequate consideration of the way it is set up, for the training of staff and for exactly how it is to be used. Implementation of this modern technology that appears to promise all, can potentially hold many dangers and conventional techniques should never be ignored at the expense of the PCR alone. A well-considered argument as to why PCR should be set up, along with considerations of the cost benefit in the short and long term, is necessary. This book is intended to give a concise overview of the practical aspects of PCR; to consider its best use in terms of laboratory practice; setting up of laboratories to perform PCR; GLP and standardisation of PCR protocols; to highlight difficulties as well as advantages in its application; to deal with more recent advances in the PCR methods and to place the PCR into context of other tests. It is hoped that readers of all levels and disciplines can be in a better position to judge the merits of the techniques and to develop the technology in a logical and efficient way. The book will be detailed in many areas as a direct help to those involved in everyday PCR. Thus, specific examples of protocols as they are being used as standard operating procedures (SOP's) in individual laboratories; examples of routine uses; latest developments and potential of PCR technologies and an outline of training material necessary for PCR technology transfer, are given. Specific emphasis is placed on the use of PCR for diagnosis of infectious diseases, aimed at diagnosticians and scientists of the developing world.

2. What is PCR?

Molecular biology has been revolutionised by PCR, a method that efficiently increases the number of DNA molecules in a logarithmic and controlled fashion. The concept of DNA amplification by PCR is simple and its impact has been extraordinary. Kary Mullis conceived PCR in 1983 and the first PCR publication appeared in 1985. Every year thereafter, the number of papers on PCR has risen exponentially

In 1989, Science magazine selected PCR as the "major scientific development" and Taq polymerase, the enzyme essential to PCR's success, as "molecule of the year". In 1993, Kary Mullis received the Nobel prize for chemistry. The advent of PCR meant that insufficiencies in the quantity of DNA were no longer a limitation in molecular biology research or diagnostic procedures. It is indeed difficult to find publications in the biological sciences that do not describe the application of PCR in some or other way. The chemistry involved in PCR depends on the complementarity (matching) of the nucleotide bases in the double-stranded DNA helix. When a molecule of DNA is sufficiently heated, the hydrogen bonds holding together the double helix are disrupted and the molecule separates or denatures into single strands.

[1] The PCR process and 5' nuclease process are covered by patents owned by Roche Molecular Systems, Inc. and F. Hoffman-la Roche, Ltd.

If the DNA solution is allowed to cool, the complementary base pairs can reform to restore the original double helix. In order to use PCR, the exact sequence of nucleotides that flank (lay on either side of) the area of interest (the target area that needs to be amplified), must be known. This is the absolute minimum data necessary before a typical PCR reaction can be used.

This data is necessary for the design of PCR primers that are 5'-3' oligonucleotides of about 20 nucleotides in length. These are designed to be complementary to the flanking sequences of the target area, as mentioned previously. Thus, the researcher has to either use previous data (known information of sequences) or, if this is unavailable, determine the sequence of these regions experimentally. The two primers (primer pair) can then be synthesized chemically and will then serve as leaders or initiators of the replication step.

The key to the replication reaction is that it is driven by a heat-stable polymerase molecule that reads a template DNA in the 3'-5' direction and synthesises a new complementary template in the 5'-3' direction, using free dideoxy nucleoside triphosphates (dNTP's = nucleotide bases) as building blocks.

2.1. PCR principles

The principles of the PCR are shown in **Figures 1-5.** These figures illustrate the fundamental aspects of the steps leading to the amplification of DNA in a stepwise PCR and introduce terms and some practical features. These can be summarised as:

2.1.1. Denaturation

When a double stranded DNA (dsDNA) molecule is heated to 94°C, the paired strands will separate (denature). This allows the primers access to the single stranded DNA (ssDNA) templates.

2.1.2. Annealing

The reaction mixture is cooled (about 50°C) to allow primers to select and bind (hybridize) to their complementary positions on the ssDNA template molecules.

2.1.3. Elongation

The ssDNA/primer solution is heated to 72°C. In the presence of the heat stable polymerase, PCR buffer, dNTP's and magnesium (Mg^{2+}) molecules, the replication procedure begins.

With each repetition of this cycle, the target is doubled and soon, after about 30 cycles, the reaction will yield in excess of 1 million copies of the target DNA fragment.

1a. Amplification target

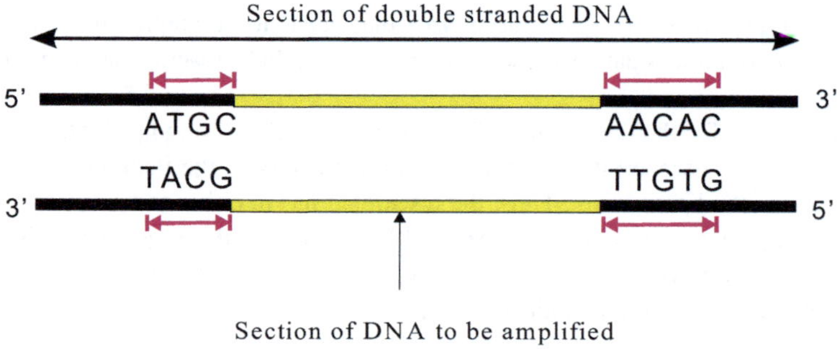

1b. PCR components

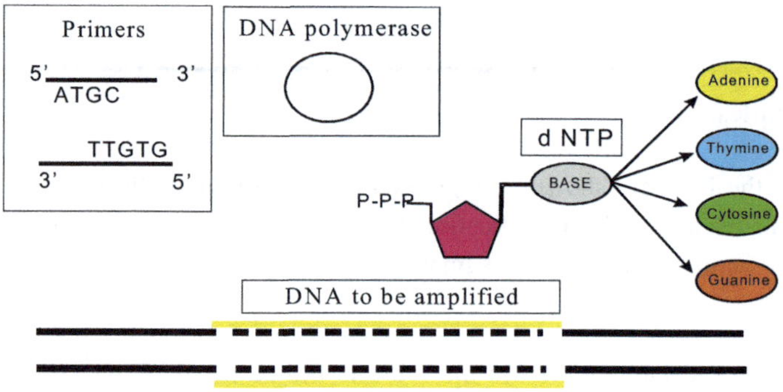

Figure 1a. Part of a double stranded DNA molecule is indicated schematically, with some bases highlighted on either side of the region to be amplified. These are the primer docking sites. Note the anti-parallel 5' to 3' orientation of the two strands.
Figure 1b. The components needed for PCR, including primers, polymerase enzyme and a mixture of bases.

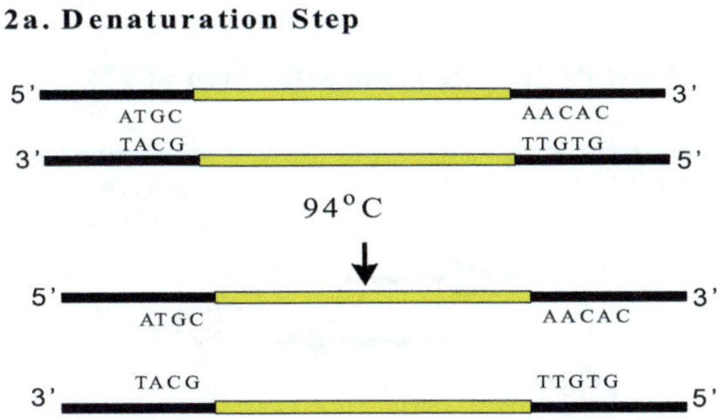

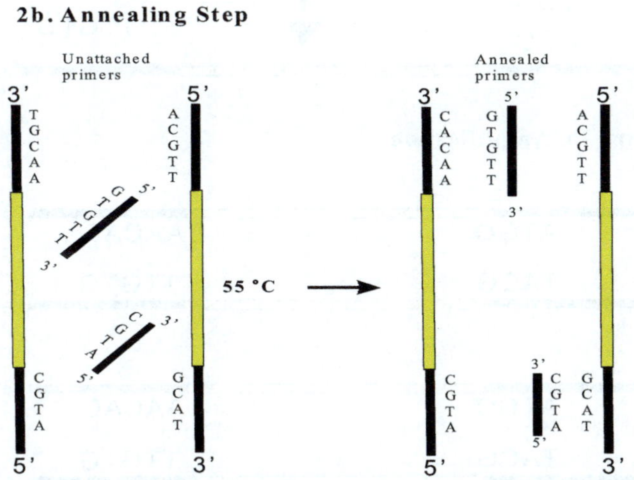

Figure 2a. The effect of heating dsDNA at 94^{0}C. Hydrogen bonds are broken and the complementary strands of dsDNA separate.
Figure 2b. Specific primers with complementary bases to ssDNA template are added. At 55^{0}C the primers anneal with the complementary ssDNA bases.

Figure 3a. First PCR cycle Elongation step at 72°

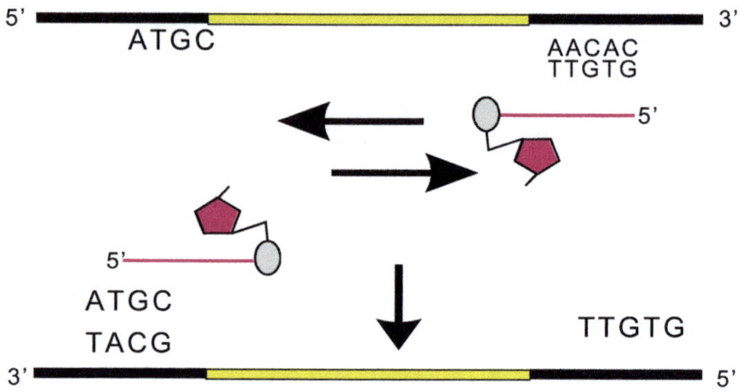

Figure 3b. Amplification

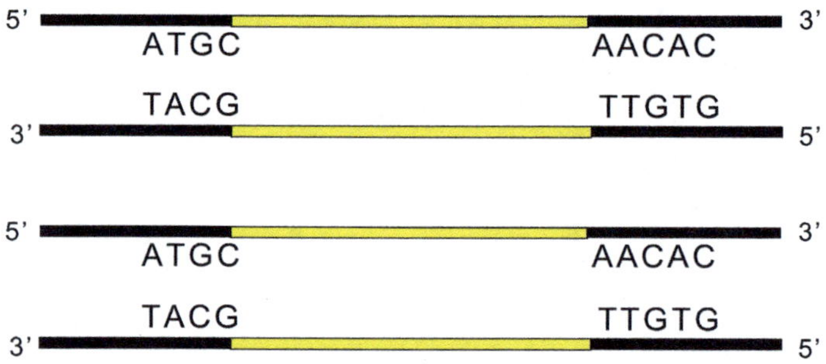

Figure 3a. Two ssDNA strands with annealed primers are indicated. The addition of a polymerase enzyme and a mixture of dNTPs allows for the production of a new strand through the addition of complementary bases to this growing strand. The DNA produced (amplicon) is derived from both template DNA strands.

Figure 3b. The target DNA segment is duplicated following a sequence of events initiated by the binding of primer pairs to the flanking regions.

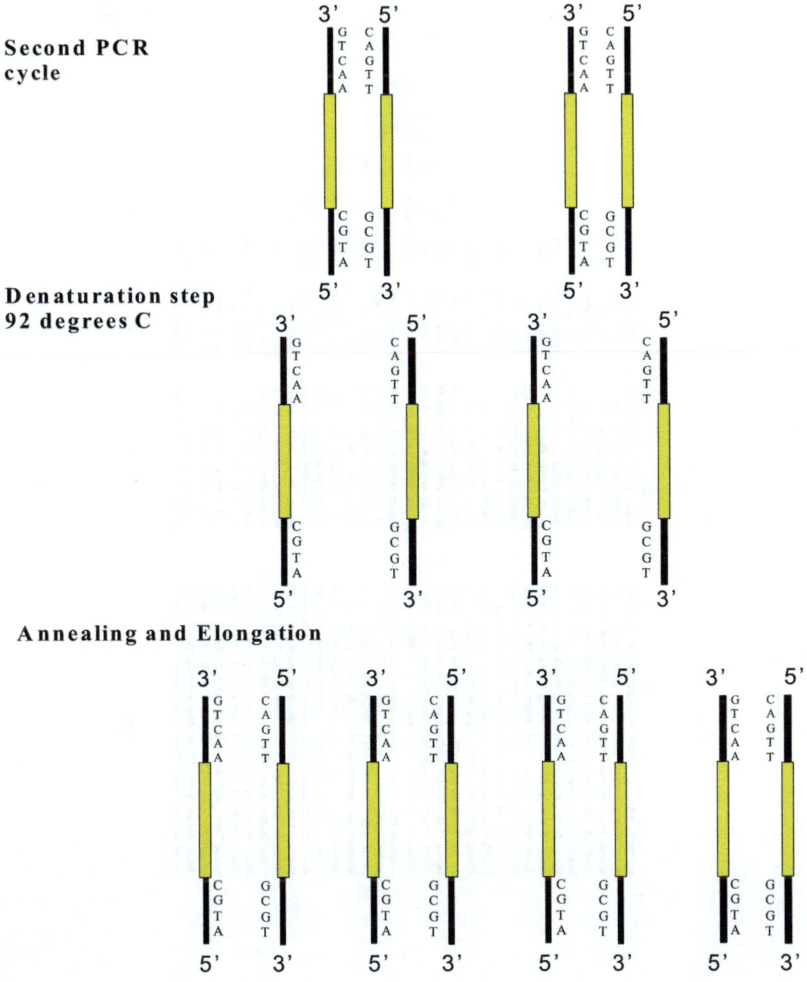

Figure 4. The second cycle of a PCR reaction: The two double strands of DNA produced (amplicon) in the first cycle are denatured to give 4 single strands. Addition of primers, enzyme and bases allows these to be copied as in the first cycle, to produce 4 double stranded amplicons.

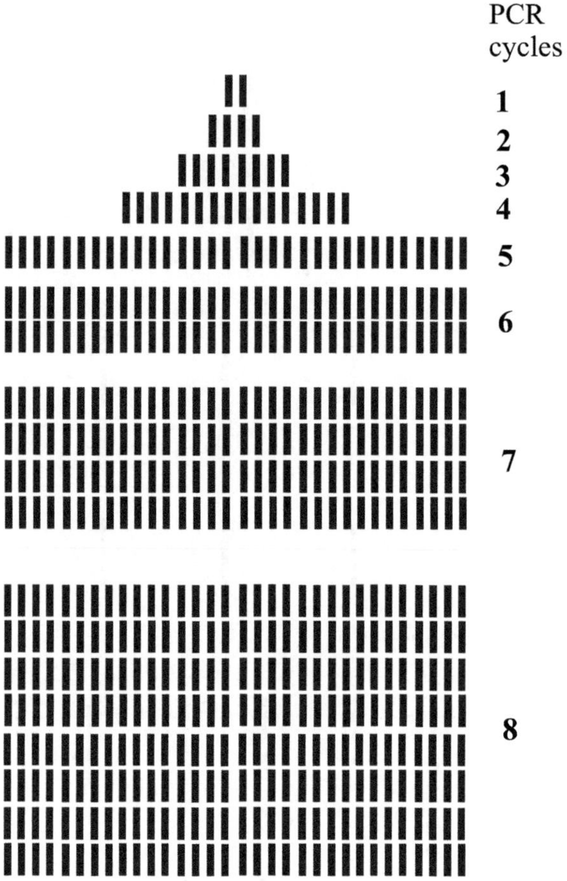

Figure 5. The increase in DNA with increased number of PCR cycles (denaturation, annealing and extension). In can be seen graphically that the number of molecules increase rapidly in a short time. In 30 cycles of PCR, a single target DNA molecule is increased to billions of copies (2^{30}).

3. What is the use of PCR?

PCR is primarily a method to spectacularly amplify a desired DNA fragment (piece of DNA) in order to increase the target DNA to detectable levels. This has had a profound effect on all molecular studies including those in the diagnostic area. It suddenly changed the way sensitivity was defined, as we are now able to detect very low numbers of pathogens with great accuracy. We are also able to detect carrier animals more easily, detect mixed populations of pathogens in an infection and to determine pathogen load. The method has found numerous related applications in molecular biology and now forms the fundamental basis of most studies involving genetic material.

As illustration of this uniqueness, PCR can be used very effectively to modify DNA. Such modification may include the addition of restriction enzyme sites (in order to facilitate cloning requirements) or regulatory elements (e.g., the addition of promoter sequences to a DNA cistron). A further type of modification can be the generation of desired site-directed mutations in a gene, inclusive of sequence alterations, additions or deletions.

Cycle-sequencing, a modification of the classical di-deoxy sequencing method pioneered by Fred Sanger in the early 1980's, uses the principles of PCR to rapidly perform sequence reactions in a thermal cycler.

The way in that PCR has dramatically impacted on diagnosis of genetic and infectious disease is one of the focus areas of this book.

For PCR-directed diagnostics, it is for example, possible to work with crude samples and minute amounts of material, that may include degraded templates, blood, sperm, tissue, individual hairs, etc.

In related applications, PCR today plays a central role in genetic typing of organisms or individuals and molecular epidemiology.

Some examples of the general application of PCR in molecular biology are given in **Table 1** and specific focus is given to the diagnostic uses of this technique in further sections of the book.

Table 1. General Applications of PCR	
1. Diagnosis of pathogens	a) PCR b) Nested PCR c) Quantitative PCR d) Multiplex PCR e) Differential on-line and real time PCR
2. Typing genetic markers	a) RFLPs b) AFLPs c) Short tandem repeat polymorphisms
3. DNA template for	a) Genomic mutation screening intron-specific primers flanking exons b) RT-PCR cDNA used as templates for pairs of exon-specific primers to generate overlapping fragments
4. Detecting point mutations	a) Restriction site polymorphisms b) Allele specific amplification
5. cDNA cloning	a) DOP-PCR b) RACE
6. Genomic DNA cloning	New members of a DNA family a) DOP-PCR. Whole genome or subgenomic amplification b) DOP-PCR c) Linker-primed PCR
7. Genome walking	a) Inverse PCR b) Bubble linker (vectored) PCR c) IRE-PCR
8. DNA templates for DNA sequencing	a) ssDNA by asymmetric PCR b) dsDNA for direct sequencing or for cloning followed by sequencing
9. *In vitro* mutagenesis	a) 5' add-on mutagenesis to create a recombinant PCR product b) Mispaired primers to change a single predetermined nucleotide

3.1. PCR and infectious diseases-the veterinary picture

Infectious diseases can be caused by microbial pathogens, include agents of fungal, protozoan, bacterial, clamydial, rickettsia and viral nature. Despite many advances in diagnostics and vaccinology, infectious diseases still have devastating consequences for agricultural, economies, worldwide.

Three examples of such devastation since the 1990s with regard to animal husbandry, include the emergence of the prion, bovine spongiform encephalopathy (BSE); the huge outbreaks of foot-and-mouth disease (FMD) in Europe and avian influenza (AI) in Asia and elsewhere.

A great many of these infectious diseases can be transmitted from vertebrate animal to man (called zoonoses) and in fact, more than 200 such zoonotic diseases are known. Infectious diseases are typically transmitted through the skin or eyes (direct contact, insect vectors, bite wounds, sexual contact). In other cases, the agents are airborne and infect the epithelial cells lining the respiratory tract from where further systemic infection may proceed. Additional sources of infectious microorganisms are contaminated food and water with a route of infection through the mouth and alimentary tract, or through the respiratory system.

Many infectious diseases can be considered new, emerging, re-emerging or resurgent. These are diseases that have been described in the last 10-30 years, or they are caused by specific modifications of agents that are already present in the environment (e.g., in a different host reservoir). These agents then evolve, mutate or are otherwise epidemiologically affected by changing conditions or other selective advantage.

Typically, re-emerging diseases are those diseases that have persisted at a subdued level in the population and recur as a result of antimicrobial drug resistance or any other changes that might favor dramatic increases in disease incidence.

Re-emerging diseases can also be described as resurgent, pertinently referring to an abrupt increase in incidence or geographic distribution of the particular disease. The emergence and re-emergence of diseases are clearly related to changes in the infectious pathogen, the vector or transmission system and the host population.

Apart from drug resistance already mentioned, some other epidemiologically important changes are: mutations that lead to increased virulence; changes in the distribution or activity of vectors; globalization and increased travel; war, population explosions; climatic and ecological changes; geographical displacement of species; movement into previously uninhabited areas; poverty and breakdown of animal or healthcare systems; changes in agriculture and industrialisation.

The quest for improved diagnostic methods in the combat against infectious diseases has become ever more demanding. As our knowledge of the structure and function of pathogens and of the immunological responses of a host to infection or antigenic stimulation expanded, approaches to both disease diagnosis and to immunisation have changed from empirical methodologies to procedures undertaken in a much more informed manner. These approaches have enabled diagnostic assays to become more specific and better standardised and have resulted in the development of vaccines of greater safety and efficacy and of a more reliable quality.

Advances in laboratory technology have enabled diagnostic assays to be performed and analysed with a greater degree of automation, with improved precision and reduced labour requirements. However, some of these advances come at an increased cost and it is therefore not always appropriate to apply them, especially with the strong budgetary constraints of most developing countries.

We now have the scientific knowledge and technical skills available to make very significant further advances in the diagnosis of, and immunisation against, specific diseases. Such expected advances will also be highlighted in the following discussion concerning laboratory diagnosis and technology.

4. Laboratory Diagnostic Technology

Laboratory diagnostic technology is directed toward either:

- The detection of the presence, or absence, of a pathogen and its subsequent identification and characterisation
- The detection of the pathological effect of, or immunological response to, infection by a particular pathogen

In the past, the detection of pathogens has been achieved by visualisation of the organism by light or electron microscopy, either directly in specimens from the affected animal or following culturing. Alternatively, serological procedures are used in that a specific antigen is detected using characterized antibodies. There have been significant advances made in such serological procedures. There have also been enormous advances in our understanding of immunology. However, with a few exceptions, immunological approaches to infectious disease diagnosis are directed towards serological procedures, i.e., the detection of interactions between antigens unique to the pathogen and specific antibodies elaborated against them. Grouping or typing of pathogens through serology has played a major role in pathogen classification and in epidemiological studies. More recently, with the advances in molecular biology, the analysis of the genetic material of pathogens has complemented or replaced serological methods for diagnostics, epidemiology and taxonomy.

Fragments of DNA or RNA produced by nuclease digestion can be separated by electrophoresis to form bands of which the position in a gel is dictated by molecular size. The patterns obtained for digests from different sources can be compared by running them in parallel. Greater discrimination can be obtained by running gels in two dimensions, under conditions producing separation in the first dimension on the basis of charge and in the second on the basis of size. Characteristic patterns are obtained (fingerprints) that can be compared visually. This can be used for comparing isolates of a particular pathogen.

There are significant advantages in the ability to indicate a pathogen's presence by the detection of its DNA or RNA. Successful bacterial or viral isolation is dependent on the presence of live or viable pathogen in a specimen and is generally time consuming and expensive. It also requires the presence of live pathogen. Antigen detection procedures are limited by the amount and quality of antigen present in specimen. Nucleic acid is more resistant to denaturation than protein and can survive long period of time (even centuries) under appropriate conditions. The limitation on nucleic acid detection has been the very small amounts that are available for detection. Notwithstanding, nucleic acid hybridisation techniques have been used to probe specimens, using a complementary strand of DNA or RNA, appropriately labelled with an enzyme or a radioisotope. Specific base pairing produces a hybrid between the probe and the target, that can be detected based on the label. Such nucleic acid probes have been developed and used for the detection of many pathogens.

PCR represents an entirely new technology. *In vitro* bacterial or viral culture is widely used to isolate and multiply a pathogen, so that the organism itself, or its antigens, can be more readily detected, by virtue of being present in greater quantity and generally with fewer contaminants. PCR technology permits the same principle (i.e., *in vitro* amplification) to be applied to the detection of specific sequences of nucleic acid. There are enormous benefits to this approach.

The application of PCR to disease diagnosis has been somewhat restricted to laboratories that have the required facilities, equipment, funding and expertise. The procedure must be undertaken in very clean conditions, since contamination with minute amounts of extraneous DNA may produce false positive results. Often this means that separate rooms and/or laminar flow cabinets must be used with careful disinfection protocols to avoid cross-contamination. Equipment costs are high but will become justifiable to more laboratories as the range of applications increase.

Methods are becoming better standardised so that training of staff in appropriate techniques can be rationalised. Certain standard reagents and consumables, such as Taq DNA polymerase, are expensive and not readily available in some countries. Nevertheless, it can be expected that these limitations will be progressively (probably rapidly) addressed and that this technology will be applied to an increasing extent.

The application of PCR (and other hybridisation techniques) to diagnostics is dependent on gene sequencing. Sequences that are unique to the pathogen must be targeted and the regions chosen for complementary primer production must be conserved within the genotypic range of the pathogen. Thus the challenge for many laboratories is access to specific primers and oligonucleotide detection probes. While these can be synthesised in suitably equipped laboratories, for most laboratories in developing countries there will be a dependence on having these reagents custom made by commercial sources or accessing them from international reference laboratories Already, there is a growing level of support from such laboratories to introduce this technology into developing countries and a few are being highlighted here.

4.1. Detection of Foot and mouth disease (FMD) virus by PCR

Antigen detection by ELISA, or CFT, is an adequate means of detecting FMD virus in epithelial samples from animals with unruptured or recently ruptured vesicles. At later stages of the disease, this becomes less reliable. For detection of virus in oesophageal/pharyngeal specimens from carrier animals, virus isolation is used, with its usual limitations. PCR was applied and described to the detection of FMD viral nucleic acid in such clinical specimens. Primers flanking a conserved region of the polymerase gene were used, so that sequence to any of the virus types could be amplified. Amplicon was detected by agarose gel electrophoresis, on the basis of molecular mass, with confirmation by hybridisation to a labelled probe. The procedure was specific for FMD viruses and at least as sensitive as alternative diagnostic procedures. Similarly, a PCR was developed directed towards the detection of FMD virus in oesophageal-pharyngeal tissues and fluids. For more detail, please refer to **Chapter 7, Section 6.**

4.2. Diagnosis of Contagious Bovine Pleuropneumonia (CBPP) by PCR

A PCR procedure has been developed to detect Mycoplasma DNA in pleural fluids from cattle with CBPP. Detailed methodology for a double PCR assay is given in **Chapter 7, Section 7.** Two assays are described as follows:

- The first assay detects a DNA sequence of about 460 base pairs from any member of the *Mycoplasma mycoides* cluster
- The second PCR is specific for Mycoplasma mycoides subspecies S.C., the agent of CBPP. It detects a 275 base pair sequence within the original sequence, with three internal restriction sites producing a characteristic pattern on agarose gel electrophoresis

This technique is highly sensitive and, combined with dot-blot hybridisation, can be used to detect as little as one colony-forming unit. This procedure is currently in use in several African national veterinary laboratories.

4.3. PCR for detecting Lumpy skin disease virus

The development of a rapid diagnostic test for lumpy skin disease (LSD) has been hindered by a number of factors. Electron microscopy is reliable but requires equipment that is not readily available in endemic areas. Also, LSD virus cannot be differentiated from another orthopoxvirus, which also produces skin lesions in buffalo. An antigen trapping ELISA is comparable to virus isolation in sensitivity but neither the ELISA nor isolation detects virus bound to neutralising antibody **(*See* Chapter 7, Section 9).** A sensitive and specific PCR was developed for the detection of capripox virus DNA in tissue culture supernatants and biopsy specimens. Two pairs of primers were used, complementary to gene sequences of the viral attachment and fusion proteins. Restriction endonuclease analysis of the products provided a simple means of confirming the identity of the PCR product. All reagents for the test are commercially available.

4.4. Differential PCR assay for rinderpest and Peste Des Petits Ruminants (PPR)

A reverse transcriptase method detects and differentiates between rinderpest and PPR viruses **(Chapter 7, Section 6).** Viral RNA is transcribed to cDNA that is then amplified using at least three primer sets. Two sets amplify regions of the fusion protein gene specific for either rinderpest or PPR viruses. The third set is based on a highly conserved region of the phosphoprotein gene that will detect rinderpest and PPR virus and other morbilliviruses. This is included to accommodate the possibility of a small change in the nucleotide sequence of the fusion protein producing a false negative result. The PCR products are resolved on an agarose gel. They are further identified by PCR using nested primer sets based on the amplified fusion protein sequence. Alternatively, the PCR products can be subjected to sequence analysis. PCR can readily be applied to detection of rinderpest and PPR viruses in tears and swabs from eyes, mouth and gum erosions of affected animals. Such samples are readily obtained and easily transported to the laboratory. Furthermore, the PCR product can be used for sequence analysis for epidemiological studies, without the need to first isolate viruses in cell culture.

4.5. Other PCR assays

A large number of specific diagnostic PCR methods have been made available in the public domain though publications in scientific journals, websites and the like. It is beyond the scope of this book to provide a comprehensive list of such methods. However, apart from the assays listed above, a few other diagnostic assays for important veterinary diseases for which established standard operating procedures (SOPs) have been described and which are widely implemented, have been included in **Chapter** 7. These include detailed methods for Newcastle disease, Bovine viral diarrhea (BVD) complex, Malignant catarrhal fever (MCF), Rift valley fever (RVF) and Brucella.

5. References

1. Armstrong, R.M., Samuel, A.R., Carpenter, W.C., Kant, R. & Knowles, N.J. 1994. A comparative study of serological and biochemical methods for strain differentiation of foot-and-mouth disease type A viruses. *Vet. Microbiol.*, 39, 285-298

2. Chamberlain, R.W., Wamwayi, H.M., Hockley, E., Shaila, M.S., Goatley, L., Knowles, N.J. & Barrett, T. 1993. Evidence for different lineages of rinderpest virus reflecting their geographic isolation. *J. Gen. Virol.*, 74, 2775-2780

3. Dedieu, L., Mady, V. & LeFevre, P.C. 1994. Development of a selective polymerase chain reaction assay for the detection of *Mycoplasma mycoides* subsp. *mycoides* S.C. (contagious bovine pleuropneumonia agent). V*et. Microbiol.*, 42, 327-339

4. Donn, A., Martin, L.A. & Donaldson, A.I. 1994. Improved detection of persistent foot-and-mouth disease infection in cattle by the polymerase chain reaction. *J. Virol. Meth.*, 49, 179-186

5. Dunn, C.S. & Donaldson, A.I. 1997. Natural adaptation to pigs of a Taiwanese isolate of foot-and-mouth disease virus. V*et. Rec.*, 141, 174-175

6. Forsyth, M.A. & Barrett, T. 1995. Evaluation of a polymerase chain reaction for the detection and characterisation of rinderpest and peste des petits ruminants for epidemiological studies. *Virus Res.*, 39, 151-163

7. Harding, M.J., Prud'homme, I., Gradil, C.M., Heckert, R.A., Riva, J., McLaurin, R., Dulac, G.C. & Vydelingum, S. 1996. Evaluation of nucleic acid amplification methods for the detection of hog cholera virus. *J. Vet. Diagn.Invest.*, 8, 414-419

8. Ireland, D.C. & Binepal, Y.S. 1998. Improved detection of capripoxvirus in biopsy samples by PCR. *J. Virol. Meth.*, 74, 1-7

9. King, A.M.Q., Underwood, B.O., McCahon, D., Newman, J.W.I. & Brown, F. 1981. Biochemical identification of viruses causing the 1981 outbreaks of foot and mouth disease in the UK. *Nature*, 293, 479-480

10. Meyer, R.F., Brown, C.C., House, C., House, J.A. & Molitor, T.W. 1991. Rapid and sensitive detection of foot and-mouth disease virus in tissues by enzymatic RNA amplification of the polymerase gene. *J. Virolog. Meth.*, 34, 161-172

11. Palmenberg, A.C. 1989. Sequence alignments of picornaviral capsid proteins. In: Molecular Aspects of Picornaviral Infection and Detection. Semler, B.L. and Ehrenfeld, E. (eds.). American Society for Microbiology, Washington, USA. pp. 211-241

12. Saiki, R.K., Gelfand, D.H., Stoffel, S., Scharf, S.J., Higuchi, R., Horn, G.T., Mullis, K.B. & Erlich, H.A. 1998. Primer-directed enzymatic amplification of DNA with a thermostable DNA polymerase. *Science*, 239, 487-491

Chapter

2

CHAPTER TWO
PCR – THE BASIC REACTION

1. Introduction

As mentioned in the previous Chapter, PCR has numerous applications in molecular biotechnology. This section briefly deals with the basic technique and the optimization of the reaction and will be a useful guide towards getting started in PCR and towards enhancing PCR's in terms of yield, specificity and fidelity. Further on, specific modifications of the basic PCR, limitations of PCR's and troubleshooting problems are discussed.

2. Basic reactions

In performing a PCR reaction, several critical constituents are needed. Firstly, the sample that contains the DNA one wishes to amplify (template DNA) is required (the synthesis of complementary DNA (cDNA) from RNA genomic templates will be discussed later). Then, two different and specific oligonucleotides (primers) that will bind to opposite strands flanking the amplification target on the template DNA, must be available.

To catalyze the amplification, a heat-stable DNA polymerase enzyme is used and this enzyme needs to be supplied with an appropriate buffer, metal ion cofactor(s) and deoxyribosenucleotide building blocks, dNTP's, **(Table 1).** Other optional components may be added for specific applications (i.e., to stabilise the enzyme, manipulate the denaturing temperatures, etc.). In a "typical" PCR reaction, the following components are mixed in a 0.2 or 0.5 mL PCR tube and made up to 25-100 µL with double distilled (dd) H_2O. The reaction mixture is overlaid with 3 drops of mineral oil (i.e., completely cover the surface of the mix with 2 mm oil layer). In the case of PCR machines with heating tops, the mineral oil is not needed.

Table 1. Basic elements of reaction mixture for PCR

Template DNA	10^5-10^6 target molecules
Primer:	0.1 µM each
10x buffer	As supplied with the enzyme
$MgCl_2$	1.5 mM (if not included in buffer)
dNTPs	100 µM of each of dATP, dCTP, dGTP and dTTP
Heat-stable DNA polymerase	1-2 Units

2.1. Temperature profile for the PCR cycles

Any PCR essentially involves a number of cycles at different temperatures. One where the template is denatured (+94°C); another where primers are annealed (hybridised) to the template (wide range of possible temperatures up to 72°C and a third temperature cycle where primers are extended (72°C). Such a temperature profile is shown in **Figure 1.**

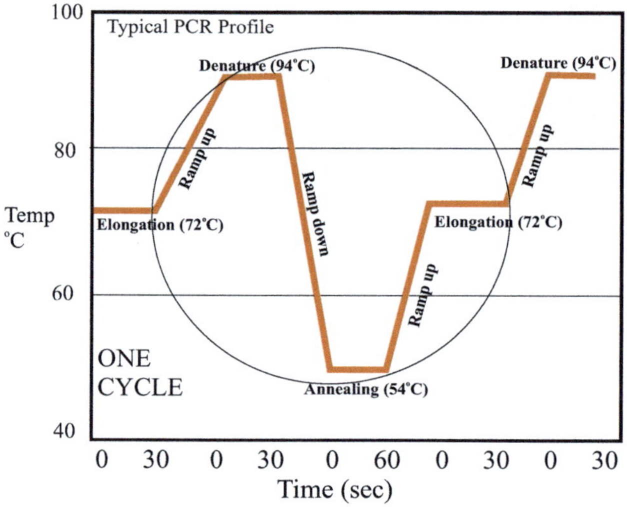

Figure 1. Schematic illustration of a typical PCR temperature profile.

In the early days of PCR, tubes were shifted by hand among three different waterbaths or heating blocks that were kept at different temperatures. However, automation has quickly followed and amplification can now conveniently be performed in a DNA Thermal Cycler. Today, a large variety of such cyclers are available (*see* **Appendix A**). Some of these also offer on-line real-time detection of the amplification reaction and will be discussed in due course. A typical temperature profile is dependent on a variety of factors such as target and primer lengths and composition and other variables that are discussed in further sections.

Table 2. Typical cycle for PCR Such a cycle is usually repeated 25-30 times	
Denaturation	94°C, 20 s (longer initial denaturation is usually required (30 s) when working with chromosomal or complex genomic material)
Primer annealing	55°C, 20-30 s
Primer extension	72°C, 1 min extension per 1000 bp amplicon
Final extension	72°C for 5 min (Especially when amplifying long products, or when cloning of the amplicon is considered)

2.2. Template DNA

For the many different applications of the PCR, many different methods of isolating and preparing the template DNA exist, mostly depending on the source of the DNA. It is important to note that the outcome of a PCR is dependent on the quality and integrity of the template DNA. It is wise to purify template DNA using a product or method that is specifically designed to purify template DNA for use in PCR **(further discussed in the protocol Sections of Chapter 7 and 8)**. The amount of input template DNA is also of crucial importance in PCR and a common mistake is to add too much template DNA to the PCR reaction. Generally, the amount of DNA per reaction should be 10^4-10^6 target/template molecules. The following guidelines **(Tables 3, 4, 5)** could serve as a rule of thumb for different types of samples from which DNA is commonly amplified.

Table 3. Guidelines for template DNA input		
Human DNA	0.5 µg	$1x10^5$ targets
Blood (human)	1 µL	$7.5x10^4$ targets (40 ng)
Guthrie blood spot	2.5 mm spot	10^5 targets
Semen	30 µL	$3x10^5$ targets (5 µg)
Yeast:	10 ng	$3x10^5$ targets
***E. coli* DNA**	1 ng	$1.5x10^5$ targets
Bacteriophage	1 plaque	10^6 targets

Table 4. A useful guideline for choosing the number of copies of template			
Number copies starting template	1 kb DNA	E. coli DNA	Human genomic DNA
10	0.01 fg	0.05 pg	36 pg
100	0.11 fg	0.56 pg	360 pg
1000	1.10 fg	5.60 pg	3.6 ng
10000	11.0 fg	56.0 pg	36 ng

Table 5. Molar conversions for nucleic acid templates			
Nucleic acid	Size	pmol/µL	Molecules/µg
1 kb DNA	1000 bp	1.52	9×10^{11}
Average mRNA	1930 b	1.67	1×10^{12}
E.coli	5×10^{6}*	3×10^{-4}*	2×10^{8}#
Human	3×10^{9}	5×10^{-7}	3×10^{5}#

* Base pairs in haploid genome.
\# For single-copy genes.

2.3. Primers

PCR primers are specific short strings of single-stranded DNA (ssDNA), known as oligodeoxyribonucleotides or oligomers.

These primers flank opposite strands on either end of the target DNA. The design of these primers is very important and it is the composition, sequence match with the template and the concentration of primers that play an important role in the outcome of a PCR assay.

2.3.1. Primer concentration

Primer concentrations should be 0.1-0.5 μM in optimal reactions of a standard or basic nature.

Higher primer concentrations (> 0.5 μM) may cause the accumulation of nonspecific products by promoting priming at nonspecific sites on the template (mispriming).

The primers should nevertheless be in excess in order to avoid their exhaustion before the completion of the reaction that would compromise the yield of the desired amplicon.

As a general rule, 5 pmol of each primer / 25 μL PCR reaction, 10 pmol/50 μL PCR reaction and 20 pmol/100 μL PCR reaction.

2.3.1.1. Calculations

In order to obtain a molarity of 0.1-0.5 μM, the corresponding molar amount of primer needed in a 100 μL reaction, is 10-50 pmol.

It is useful to dilute oligonucleotide primer stocks to a concentration of 10 pmol/μL (10 μM), enabling the addition of a convenient volume of primer stock to the PCR reaction (1-5 uL/100 μL PCR reaction).

After synthesis, the concentration of the oligonucleotide primer may be expressed in different units, depending on the manufacturer (usually in optical density units (OD), and/or in μg). The following approximate values are useful in calculating your required dilution of the concentrated primer solution.

2.3.1.2. Formulae

1 pmol of a 30 mer oligo	=	10.26 ng
1 μg of a 30 mer oligo	=	0.0975 nmol
1 OD_{260} (ssDNA)	=	33 μg/mL

2.3.1.3. Example 1

G (+) is a 23 mer PCR primer
Optical density is measured as 77 OD_{260}

You need to use 10 pmol/100 μL PCR reaction (suitable primer dilution is 10 pmol/μL).
Using the above approximate values, the calculation may be:

Formula:	1 pmol of a 30 mer oligo = 10.26ng
Therefore:	1 pmol of a 23 mer = 7.87 ng (i.e., less weight!)
Required dilution:	10 pmol/μL = 78.7 ng/μL = 78.7μg/mL

i.e. 77 OD x [OD_{260} for ssDNA] = 77 x 33μg/mL = 2541 μg/mL.

$(M_1V_1) = (M_2V_2)$; 2541 μg/mL x Z μL = 78.7 μg/mL x 1000 μL. Z = 30.9 μL

Therefore: 30.9 μL of concentrated primer made up to 1 mL
will yield a concentration of 78.7 μg/mL = 10 pmol/ μL

Short formula

a/30 x 10.26/33 x 10000/b = μL amount of primer concentrate that should be made up to 1mL to yield 10 pmol/μL (where a is the primer length and b is the primer concentration in OD/mL).

2.3.1.4. Example 2

L(-) is a 24 mer PCR primer Concentration is measured as 3.39 μg/μL.

You need to use 10 pmol/100 μL PCR reaction.
Suitable primer dilution is 10 pmol/μL.
Using the above approximate values, the calculation may be:

Formula	1 μg of a 30 mer = 0.0975 μMol
Therefore	1 μg of a 24 mer = 0.122 μMol (i.e., more moles!)
Therefore	3.39 μg/uL gives 0.122 x 3.39 = 0.41 μMol/μL = 410 nMol/μL and 1 μL will give 410 nMol of primer

A 41 x dilution will yield 10 pMol/uL:
e.g., 20 μL of the concentrated primer made up to 820 μL total volume.

Short formula

30/a x 0.0975 x b x 100 = dilution needed to yield 10 pMol/ µL (where a is the primer length and b is the primer concentration in µg/µL)

2.3.1.5. Example 3.

When you order primers from a commercial company they will ask you the following information.

- Sequence (5'-3') of the primer, number of nucleotides
- The Unit size (e.g., 100 nol)
- Purification standard (e.g., standard desalting or gel purified)

The primer will then be synthesized and sent to you in a lyophilised (freeze-dried) form, together with the following data.

- Molecular weight
- GC content
- Tm (50 mM NaCl)
- Amount of oligo (lyophilised). This could be expressed as OD_{260}, nM and µg

Stock solution

Take for example a primer data sheet with the following information:

- OD_{260} = 11.7
- Molarity = 51.88nM
- Weight = 390 µg

If you make this primer up to 1mL you have 51.88 nmol/mL or 51.88 Mol/µL

Work solution: if you then dilute the stock solution 1:5, you will have 10 pMol/µL

2.3.2. Design of primers

- The design of primers, more than anything else, will determine the success of a specific PCR assay. The two primers are unrelated to one another, since they anneal to different strands and on opposite ends of the target amplicon.

- However, special care must be taken to ensure that there is no significant complementarity within or between primers and that primer pairs are balanced with respect to the melting temperature (Tm).

- When two single-stranded DNA molecules hybridise, such as in the annealing of a PCR primer to the template, the stability of the duplex is dependent on the sequence and number of associated basepairs, the concentration of the DNA species, and the salt concentration of the solution.

- The strands of the duplex can be separated by heat and the temperature at which half the molecules are single-stranded and half are double-stranded, is called the Tm. Thus, the Tm will determine the annealing temperature (Ta) in the cycler temperature profile and should suit both primers.

- A number of other pointers are also of importance in the design of primers that will function optimally, and will be discussed briefly. All of these criteria can be met in most cases by careful analysis and thought, but for convenience, several primer design software programs are available with most software packages for general DNA sequence analysis.

2.3.2.1. Additional information

Primers may contain 5' extensions or mismatches in order to engineer specific domains into the PCR amplicon. These may be required for a large variety of applications and may include:

- Restriction enzyme sites
- $^{-3}$CCATGG^{+4} (Kozak sequence to increase translation)
- Promotor sequences, enhancers and other regulatory sequences
- Internal mismatches for mutagenesis

	Table 6. Features of Primers
a)	Typical PCR primers are anything between 18-28 nucleotides in length
b)	The G+C composition should ideally be similar to that of the desired amplicon and should in general be between 50-60%
c)	The calculated Tm for a primer pair should be balanced
d)	Rule of thumb: Tm = 2(A+T) + 4(G+C) and -1.5 °C for every mismatch
e)	A Tm 55°C -72°C is desired (62-65°C is best)
f)	Check for complementarity in 3' ends of primer pairs - this lead to primer - dimer artifacts
g)	Avoid any significant secondary structure within primers i.e. internal palindromic sequences
h)	Runs of 3 or more C's and G's at 3' ends promote mispriming in G/C rich regions
i)	Palindromic sequences within the primers should be avoided
j)	Avoid an A and especially a T at the 3' end of a primer (this allow 'breathing' in the hybridisation of the primer to the template)
k)	Avoid any potential mismatches in the 3'end of primers

2.3.3. Primer annealing

Applicable annealing temperatures (Ta) are typically 5°C below the true Tm, but optimal annealing temperatures are often higher (5-10°C) than the Tm of the primers.

This optimal annealing temperature is also externally dependent on the salt concentration as well as the Mg^{2+} concentration in the reaction and has to be determined empirically.

The highest possible annealing temperatures permitted by a specific primer set should be selected since increasing annealing temperatures enhances discrimination and reduces mis-extension.

Thus a high stringency in the annealing temperature will ensure a good quality product and this should typically be in the range of 55-72°C. Annealing will require only a few seconds at the correct primer concentrations.

2.3.4. Primer extension

The time required for extension of the primers and synthesis of the entire length of the target amplicon, is dependent upon the length and concentration of the target sequence and upon the temperature and the properties of the specific enzyme used. The speed of the cycler and the properties of the tubes used may also be influential, as will later be discussed.

- Extension temperature of 72°C is near the optimal working temperature of most of the heat-stable polymerase enzymes
- Rates of incorporation varies between 35-100 nucleotides/second and therefore, as a general indicator, 1 min at 72°C is sufficient for 2kb products
- Longer cycles may be useful, especially early on if substrate concentration is very low and later on when product concentration exceeds enzyme concentration. A long last cycle (e.g., 10 min at 72°C) is useful to ensure that all amplified copies are full-length

2.3.5. Degenerate PCR primers

Occasionally, the exact nucleotide sequence of the target-template DNA will not be known and the sequence has to be deduced from the amino-acid sequence. To enable such templates to be amplified by PCR, degenerate primers can be used. These are actually mixtures of several primers whose sequences differ at the positions that correspond to the uncertainties in the template sequence.

2.3.6. Design and use guidelines

- Avoid degeneracies in the last 3 nucleotides at the 3' end of the primer
- If possible, use Met or Trp-encoding triplets at the 3' end
- To increase primer-template binding efficiency, reduce degeneracy by allowing mismatches between the primer and template, especially towards the 5' end (thus allow primer annealing at lower than optimal Tm). Again, avoid mismatching at the 3' end
- Design primers with less than 4-fold degeneracy at any given position.
- Increase the primer concentration to 1 μM

2.3.7. Storage of primers

- Oligonucleotides are chemically stable. Left dry, they should be good for years. Once hydrated, they are susceptible to degradation by nucleases. If handled correctly, they should be stable for years

- Any DNA oligonucleotide can be degraded by microbial or fingertip nucleases

- Primers should be stored in deionized water or buffered solutions containing EDTA and kept frozen when not in use

- Make a concentrated stock solution in water or TE (10 mM Tris pH8.0, 1 mM EDTA). A convenient stock could be 100 µM, stored at -20°C

- We advise that stock solutions be distributed into several tubes for long term storage so that accidental contamination of a tube will not lead to loss of the entire synthesis

2.3.8. Example primer formulation

1. **Make your lyophilised primer up to a convenient volume to contain 50 nmol of primer (*see* Section 2.3.1.5), example 3)**

2. Dissolve 50 nmol of primer in 500 µL TE to make a 100 µM stock solution

3. Dilute from this stock 1:10 in sterile water to make a working solution at 10 µM for use in the PCR reactions

4. Adding 1 µL of the 10 µM primer to a 20 µL PCR reaction will result in a final primer concentration of 0.5 µM, or 10 pMol of primer in 20 µL volume

2.4. Buffer and components

- A buffer that provides ionic strength and buffering capacity during the reaction is required and these aspects of the ionic environment of the PCR are critical
- The standard buffer is usually 10-50 mM Tris-HCl (pH 8.3- 8.8), and up to 50 mM KCl may be included to facilitate primer annealing
- Other buffers used in specific applications may include NaCl (allows better denaturation of GC rich templates) or ammonium sulphate (may limit artifacts due to incomplete amplicons)
- The addition of DMSO can be useful when amplifying multiple sequences in the same reaction (reduce secondary structure of the target DNA), but is generally not recommended because of the inhibitory effect on the enzyme
- Gelatine, bovine serum albumin (BSA), Formamide or Triton X-100 may be included in the enzyme storage buffer by some manufacturers to stabilise the enzyme

2.5. Magnesium concentration

The concentration of the essential co-factor, $MgCl_2$, can have a particularly profound effect on the specificity and yield of the PCR.

- The Mg^{2+} concentration affects the reaction differently at low or high concentrations and is influential in terms of specificity and yield of the amplification reaction
- Mg^{2+} forms soluble complexes with the dNTP building blocks to make them available and recognisable as substrate for the enzyme
- The concentration of free Mg^{2+} is determined by the presence and concentration of chelating agents like EDTA, free pyrophosphates, and dNTP's

The following general factors should be kept in mind:

- It is best to determine the optimal Mg^{2+} concentration empirically for each PCR assay
- PCR's should contain 0.5-2.5 mM Mg^{2+} and the most commonly used Mg^{2+} concentration is 1.5 mM, when the four dNTP's are present at 200 μM each) DNA, dNTPs and proteins will bind to Mg^{2+}
- Total $[Mg^{2+}]$ = (bound + free) $[Mg^{2+}]$
- Free $[Mg^{2+}]$ should equal the [dNTP]
- Presence of EDTA, hemoglobin, heparin and other chelators in the PCR reaction mixture may influence the apparent $[Mg^{2+}]$
- It is preferred that blood samples be collected in the presence of citrate. The second best is EDTA blood-tubes, but heparin tubes should never be used
- Self-adjusting magnesium buffers that automatically adjust the Mg^{2+} concentration throughout the PCR reaction can also be used. The so called 'reformulated' Mg^{2+} concentration is 2.0mM Mg^{2+} can be used if required.

Magnesium concentration may influence

- Primer annealing
- Strand dissociation temperatures
- Product specificity
- Formation of primer-dimer artefacts
- Enzyme activity/fidelity

2.6. dNTP's

A working stock containing 10 mM of each dNTP (pH7.0) is recommended and a final reaction concentration between 20 and 200 μM each (mostly depending on amplicon size, [Mg^{2+}]), result in the optimal balance in yield, specificity and accuracy. This molarity represent a sufficient excess to allow for the concentration of dNTP's to remain virtually constant throughout the reaction. Higher dNTP concentrations lead to a decrease in the specificity and fidelity of PCR. It is important that dNTP'S should be used at equivalent concentrations (i.e., should be balanced) in order to minimize misincorporation errors. In general: 200 μM of each dNTP/2 kbp amplicon @ 30 amplification cycles.

2.7. Enzymes and enzyme concentration

The recommended polymerase concentration is usually 1-2.5 Units per 100 μL reaction volume (most often supplied as 5U/μL). This would represent the comparatively low molar equivalent of 1-2.5 nM, relative to all the other reaction components. This limiting concentration ensures fidelity since higher enzyme concentrations result in nonspecific background products. When using other additives in reactions, their effects on the enzyme should be kept in mind. Glycerol for example increases thermal stability of the enzyme but reduces the Tm. DMSO also lowers Tm but reduces thermal stability of the enzyme (reduced Tm allows lower temperature and therefore less DNA damage). The polymerase is a complex molecule with three activities; a 5'-3' polymerase activity, a 3'-5' exonuclease activity (proofreading activity) and a 5'-3' exonuclease activity. These activities of the polymerase enzyme are shown in **Figures 2 (a-c).**

2.7.1. Polymerase fidelity

Polymerase fidelity is influenced by multiple factors, including:

- The tendency of a polymerase to insert the wrong nucleotide
- The presence of a proofreading 3'-5' exonuclease activity that can remove mismatches
- The ease with which mismatches can be extended

Classical DNA polymerases like T4, T7 and Klenow polymerases have 3'-5' proof reading activity, but the original Taq DNA polymerase does not have such capability (***see* Table 7 for a comparison of polymerase fidelity).** However, several different types of thermostable enzymes, including some with 3-5' proofreading capabilities are now available. Some examples of different enzymes commercialised by Roche Diagnostics GmbH, Mannheim, are shown in **Tables 7, 8 and 9.**

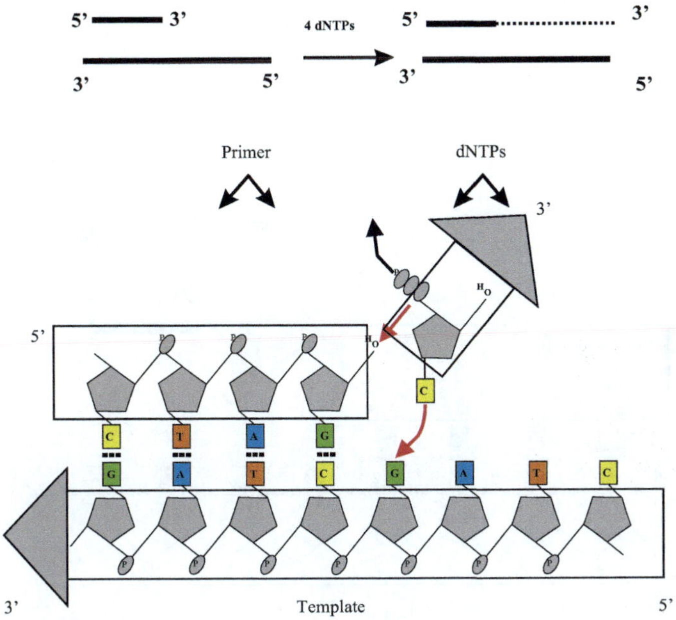

Figure 2a) DNA polymerase 5'-3' polymerase activity
Mode of polymerization 5' to 3' direction (i.e., reading complimentary strand in 3' to 5' direction from priming site

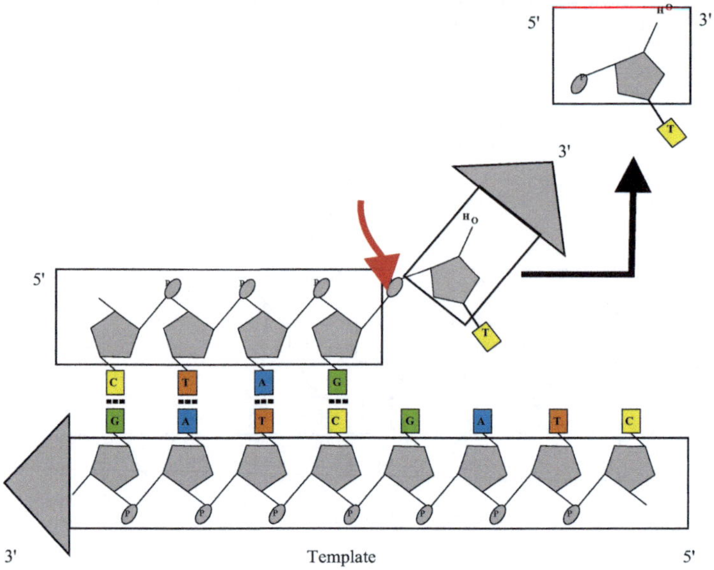

Figure 2.b) DNA Polymerase 3' – 5' exonuclease activity

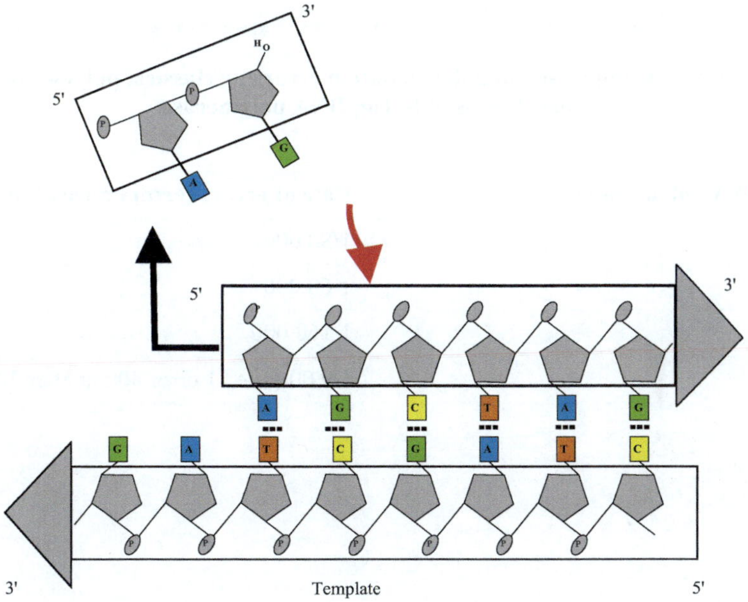

Figure 2c) DNA Polymerase 5' - 3' exonuclease activity

Table 7. Comparison of polymerisation errors by classical polymerases and heat -stable Taq DNA polymerase

Type of DNA polymerase	Rate of errors (errors/nucleotide)
T4	1/80 000
T7	1/60 000
Klenow	1/160 000
Taq	1/9 000 (i.e., 1 error/400 nt after 30 cycles)

Table 8. Source of polymerases

1)	Taq DNA polymerase was originally isolated from the thermophilic bacterium *Thermus aquaticus* BM and is now mostly supplied as a recombinant protein, produced in *E. coli*
2)	The Pwo DNA polymerase is also supplied as recombinant *E. coli* protein, and was cloned from the hyperthermophilic bacterium *Pyrococcus woesi* (an arhaeobacterium)
3)	The Tth DNA polymerase is isolated and purified from the thermophilicbacterium *Thermus thermophilus spec*

Many other thermo-stable enzymes are discovered regularly. The thermo-stable enzymes, although tolerant to high temperatures, have a limited active half-life dictated by the time that they were exposed to high temperatures.

Table 9. Comparison of PCR Enzymes

DNA polymerase	**5'-3' exonuclease**	**3'-5' exonuclease**	**Processivity (bp)** (i)	**Amplicon length (kb)** (ii)	**Error rate (10^{-6})** (iii)	Performance Long template PCR			
						To 2	2-5	3-15	10-30
Taq	+	-	50	3	26	++	(+)	-	-
Pwo	-	+	20-30	3	3.2	++	(+)	-	-
Tth	+	-	50	3	30	++	(+)	-	-

i. Processivity: Number of bp that can be added to a copy by one molecule of polymerase before it falls off the template
ii. Amplicon length. PCR fragment length that can be amplified from a human genomic DNA template with good yield
iii. Error rate per bp calculated by the Lacz assay. (***See* reference 1 for details)**

2.8. Denaturation time and temperature

The DNA strands of the amplicon template will usually fully dissociate within 20-30 min at 94-95°C, where this denaturation time is proportional to the temperature.

It may under specific conditions (e.g., very high G-C rich template) be appropriate to use higher denaturation temperatures in order to ensure complete denaturation and avoid speedy DNA renaturation or "snap-back".

One important consideration when selecting the denaturation time and temperature is however, the limited half life of the Taq polymerase at these elevated temperatures. Typically the half-life of these types of enzymes is approximately as follows:

- 72h (92.5°C)
- 40 min (95°C)
- 5 min (97.5°C)

2.9. Cycles (@)

Quote: *"If you have to go more than 40 cycles to amplify a single copy gene, there is something seriously wrong with your PCR"* - Kary Mullis.

Indeed, too many cycles increase the amount and complexity of non-specific background products. Typically the optimal number of cycles that would ensure a good PCR is between 25 and 30. In specific exceptional cases (e.g., a very small number of target DNA molecules), the number of cycles is often increased to 35 or higher.
A summary of the amplicon output at different numbers of cycles is given in **Table 10**.

Table 10. Theoretical amplicon output after different numbers of PCR cycles

Number of cycles	Copies input	Amplicon output
1	1	2
10	1	10^3
20	1	$1x10^6$
30	1	$1x10^9$

After 30 cycles one would theorectically expect $1x10^9$ copies but in reality only about $1x10^6$ copies are generated because the process is not 100% effective (see section 11 below).
Regarding input copy number, the following provides a general guideline for the number of cycles necessary for a particular PCR reaction

- < 100 input copies: 50 @
- 1000 input copies: 40 @
- >1000 input copies: 25-30 @

2.10. The Plateau effect

In order to obtain the best PCR performance in a given system:

1. Reaction component concentrations should be optimised
2. Cycle times should be optimised
3. Temperatures should be optimised

However, even when the reaction is fine-tuned, the logarithmic accumulation of amplicons is possible for a limited number of cycles only. The decrease in this exponential rate of product accumulation is known as the plateau effect (**Figure 3**). The application of PCR is in the logarithmic phase where the constant increase (2, 4, 8, 16, 32) of the amplicon number is evident (i.e., the power of PCR). Where the increase of the amplicon copy number is not in the logarithmic phase any more (due to inhibition or reagent limitations), the accumulation of amplicons enter the stationary phase or plateau. The following factors may influence plateau.

- Utilisation and depletion of substrates (dNTP, primers, template)
- Reduction in denaturation efficiency
- Inefficiency in primer annealing
- Stability of reactants (dNTP, enzyme), thermal inactivation and limiting enzyme concentration.
- End product inhibition (pyrophosphate, duplex DNA)
- Product re-association
- Competition for reactants by non-specific products, non-specific products may continue to amplify preferentially

The plateau is inherent to PCR and cannot be avoided. The substrate excess, one of the major reasons for the onset of the linear/stationary phase is, fortunately, also an indicator of the success of the PCR. Multiple reactions can be set up in cases where a large amount of PCR amplicon is desired.

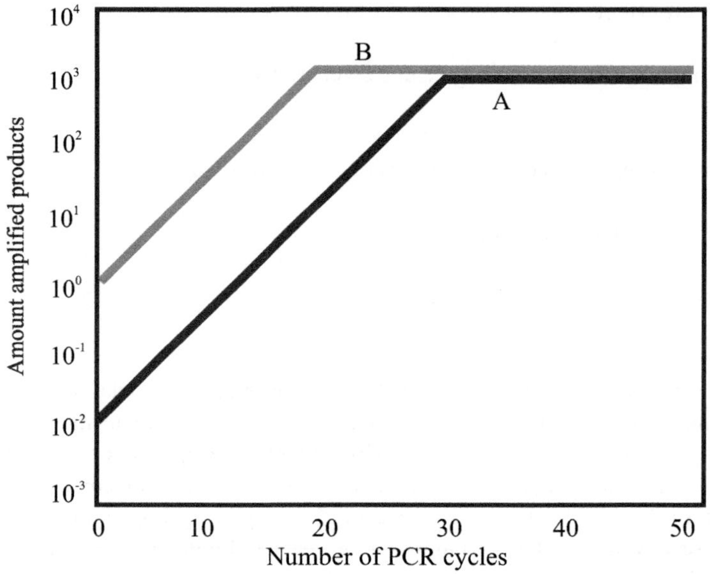

Figure 3. The plateau effect in PCR.

Despite very different amounts of input template DNA, samples A and B will yield almost the same amount of product after about 30 cycles of PCR.

3. Optimising a PCR application

Optimal PCR conditions can be very different for different assays, as influenced by the unique template, conditions and primer sets, other biochemical variables and the specific type of cycler used. When setting up a new PCR assay, a simple block titration may be designed to optimise the PCR with respect to template, Mg^{2+} concentration, primer concentrations and annealing temperatures, enzyme concentration, optimal reaction pH, the effect of additives to the reaction and any other variable that may influence the reaction.

Some commercial companies also offer a PCR optimisation kit that may be used to simplify the PCR optimisation procedure. In addition, approaches such as the touchdown PCR also offers simple one-step optimization of PCR reactions that are expected to be sub-optimal with regard to primer/template homology. As a general rule however, any PCR that will become an established assay in the laboratory should be properly optimised by a titration method. We like to recommend a most useful simplified protocol for PCR optimisation, based on a set of methods that are widely applied development trials for industrial process design, the so-called Taguchi methods.

3.1. Taguchi Method for PCR optimisation

With the use of Taguchi methods, the range of optimal conditions for any given PCR assay may be obtained in a fraction of the number of reactions required to optimise several different variables in a classical block titration.

If for example 4 different variables (dNTP, Mg^{2+}, primer, enzyme) are tested at 3 different concentrations each, the titration would need 81 reactions.

However, using Taguchi methods, only 9 reactions are required.

In this method, the number of reactions (E) is calculated from $E = 2k + 1$; with k being the number of variables to be tested. With an increasing number of variables to be tested, the reduction in the number of reactions to be performed become more drastic. For 8 variables, 6561 reactions will have to be tested in block titration, but only 17 reactions are needed with the use of the Taguchi method.

In the Taguchi method, the variables to be tested are arranged into an orthogonal array **(Table 11).** In this array, each column would represent one of the variables and each row represents one specific reaction (PCR tube):

Table 11. Taguchi array for PCR				
Reaction (tube)	**Variable 1**	**Variable 2**	**Variable 3**	**Variable 4**
1	A	A	A	A
2	A	B	B	B
3	A	C	C	C
4	B	A	B	C
5	B	B	C	A
6	B	C	A	B
7	C	A	C	B
8	C	B	A	C
9	C	C	B	A

Each of the variable components will now be added to the reaction at one of the three different concentrations selected. The three concentration values should be selected so as to fall within the range of concentrations that may be most influential on the reaction, and should be sufficiently separated to allow for the determination of their effects on the reaction. The following is an example of such selected concentrations for a given PCR assay.

Table. 12. Example of variables and concentrations to be tested in the optimization of a PCR assay			
	A	**B**	**C**
Primer: (pM)	7.5	15	30
$MgCl_2$: (mM)	0.5	1.5	3
dNTP's (µM)	50	100	200
Enzyme (U)	0.5	2	5

Using this method, the appropriate number of reactions may be assembled according to the Taguchi array **(example in Table 11)** and using the preselected variables and concentrations (for the specific PCR assay. The most common variables that should usually be optimized first are:

- Annealing temperature
- Mg^{2+} concentration
- Primer concentration
- Template concentration

This method for initial optimisation of PCR reactions offers flexibility in optimising the interaction between key components of the reaction and may be altered and fine-tuned to suit any specific PCR application.

4. References

1. Bell, D.A., DeMarini, D.M. 1991. Excessive cycling converts PCR products to random-length higher molecular weight fragments. *Nucl Acids Res.*, 19, 5079

2. Breslauer, K.J., Frank, R., Blbcker, H., Marky, L.A. 1986. Predicting DNA duplex stability from the base sequence. *Proc Natl Acad Sci USA*, 83, 3746-3750

3. Cheng S. 1994. Long PCR. *Nature*, 369, 684-685

4. Cheng S. 1995. Template Integrity is Essential for PCR Amplification of 20- to 30-kb Sequences from Genomic DNA. *PCR Methods and Applications*, 4(5), 294-298

5. Cheng, S. 1994. Effective amplification of long targets from cloned inserts and human genomic DNA. *PNAS*, 91, 5695-5699

6. Chou, Q., Russell, M., Birch, D.E., Raymond, J., Bloch, W. 1992. Prevention of pre-PCR mis-priming and primer dimerization improves low-copy-number amplifications. *Nucl Acids Res.*, 20, 1717-1723

7. Cobb, B.D., Clarkson, J.M. 1994. A simple procedure for optimizing the polymerase chain reaction using modified Taguchi methods *Nucl Acids Res.*, 22(18): 3801-3805

8. D'Aquila, R.T., Bechtel, L.J., Videler, J.A., Eron, J.J., Gorczyca, P., Kaplan, J.C. 1991. Maximizing sensitivity and specificity of PCR preamplification heating. *Nucl Acids Res.,* 19, 3749

9. Demmler, G.J., Buffone, G.J., Schimbor, C.M., May, R.A. 1988. Detection of cytomegalovirus in urine from newborns by using polymerase chain reaction DNA amplification. *J Infect Dis.,* 158, 1177-1184

10. Eckert, K.A., Kunkel, T.A. 1991. The fidelity of DNA polymerases used in the polymerase chain reactions. In PCR A Practical Approach; Eds. Mcpherson M.J, Quirke P, Taylor GR, *IRL Press*

11. Erlich, H.A., Gelfand, D., Sninsky, J.J. 1991. Recent advances in the polymerase chain reaction. *Science*, 252, 1643-1651

12. Foord, O. 1994. Long-distance PCR. *PCR Methods and Applications,* 3(6), S149-S161

13. Fujiwara, H. 1995. PCR with Deoxyinosine-containing Primers Using DNA Polymerases with Proofreading Activity. *PCR Methods and Applications*, 4(4), 239-240

14. Finckh. U., Lingenfelter, P.A., Myerson, D. 1991. Producing single-stranded DNA probes with the *Taq* DNA polymerase: A high yield protocol. *BioTechniques*, 10, 35-39

15. Gelfand, D.H. Taq DNA polymerase. 1989. In PCR Technology Principles and Applications for DNA Amplification; Ed. Erlich H.A., *Stocktom Press*

16. Gelfand, D.H. and White, T.J. 1990.Thermostable DNA polymerases. In PCR Protocols A Guide to Methods and Applications; Eds. Innis, M.A., Gelfand, D.H., Sninsky, J.J., White, T.J., *Academic Press*.

17. Her, C. 1995. Rapid Restriction Mapping by use of long PCR. *BioTechniques*, 19(4), 530-532

18. Holland, P.M., Abramson, R.D., Watson, R., Gelfand, D.H. 1991. Detection of specific polymerase chain reaction product by utilizing the 5' -> 3' exonuclease activity of *Thermus aquaticus* DNA polymerase. *Proc Natl Acad Sci., USA*, 88, 7276-7280

19. Hoof van Huijsduijnen, R.A.M., Ayala, G., DeLamarter, F. 1992. A means to reduce the complexity of oligonucleotides encoding degenerate peptides. *Nucl Acids Res.*, 20, 919

20. Innis, M.A., Gelfand, D.H. & Sninsky, J.J. (eds), 1995. PCR Strategies. Academic Press, Inc, California, USA

21. Innis, M.A., Mayambo, K.B., Gelfand, D.H., Brow, M.A.D. 1988. DNA sequencing with *Thermus aquaticus DNA* polymerase and direct sequencing of polymerase chain reaction-amplified DNA. *Proc Natl Acad Sci USA*, 85, 9436-9440

22. Jones, D.H., Winistorfer, C. 1992. Sequence specific generation of a DNA panhandle permits PCR amplification of unknown flanking DNA. *Nucl Acids Res.*, 20, 595-600

23. Koike, K., Urata, Y., Koike, M. Human Pyruvate Dehydrogenase beta-subunit-gene, complete cds. Un published data, EMBL accession number D90086

24. Kwok, S. 1991. A Guide to the Design and Use of Mismatched and Degenerate Primers. *PCR Methods and Applications*, 3(4), S39-S47

25. Kwok, S., Kellogg, D.E., McKinney, N., Spasic, D., Goda, L., Levenson, C., Sninsky, J.J. 1990. Effects of primer-template mismatches on the polymerase chain reaction: human immunodeficiency virus type I model studies. *Nucl Acids Res.*, 18, 999-1005

26. Lawyer, F.C., Stoffel, S., Saiki, R.K., Mayambo, K., Drummond, R., Gelfand, D.H. 1989. Isolation, characterization, and expression in *Escherichia coli* of the DNA polymerase gene from *Thermus aquaticus. J. Biol Chem.*, 246, 6427-6437

27. McPherson, T. PCR with highly degenerate primers *PCR: A Practical approach. Ed. McPherson. 1991. New York: Oxford University,* pp. 171-186

28. Mertz, C.E. 1994. Nucleotide imbalance and PCR: Effects on DNA amplification and synthesis of high specific activity radiolabelled DNA probes. *Anal. Biochem.*, 221, 160-165

29. Mezei, L.M. 1990. Effect of oil overlay on PCR amplification. *Amplifications A Forum for PCR Users*, 4, 11 -13

30. Mullis, K.B. 1991. The polymerase chain reaction in an anemic mode: How to avoid cold oligodeoxyribonuclear fusion. *PCR Methods and Applications*, 1, 1-4

31. Mullis, K.B., Faloona, F.A. 1987. Specific synthesis of DNA in *vitro* via a polymerase-catalyzed chain reaction. *Methods Enzymol.*, 155, 335-350

32. Peterson, D.W. 1993. Cross-Species Polymerase Chain Reaction: Cloning of TATA Box-Binding Proteins. *Methods in Enzymology: Anonymous Academic Press*, pp. 493-507

33. Rappolee, D.A. 1990. Optimizing the sensitivity of RT-PCR. *Amplifications A Forum for PCR Users*, 4, 5-7

34. Reischl, U., Wittwer, C. & Cockerill, F. (eds), 2002. Rapid cycle real-time PCR. Methods and applications. Springer-Verlag Berlin Heidelberg, Germany

35. Roux, K. 1994. Using mismarched Primer-Template Pairs in Touchdown PCR. *BioTechniques,* 16(5), 812-814

36. Roux, K. 1995. Optimization and troubleshooting in PCR. PCR Methods and Applications. 4(5), S185-S194

37. Rolfs, A., Schuller, I., Finckh, U. & Weber-Rolfs, I. (eds), 1992. PCR: Clinical Diagnostics and research. Springer-Verlag Berlin Heidelberg, Germany

38. Ruano, G., Brash, D.E., Kidd, K.K. 1991. PCR: The first few cycles. *Amplifications A Forum for PCR Users,* 7, 1-4

39. Rychlik, W. 1995. Priming efficiency in PCR. *BioTechniques*, 18(1), 84-90

40. Rychlik, W., Spencer, W.J. Rhoads, R.E. 1990. Optimization of the annealing temperature for DNA amplification in *vitro. Nucl Acids Res.*, 18, 6409-6412

41. Rychlik, W., Rhoads, R.E. 1989. A computer program for choosing optimal oligonucleotides for filter hybridisation, sequencing and in *vitro* amplification of DNA. *Nucl Acids Res.*, 17, 8543-8551

42. Saiki, R.K., Gelfand, D.H., Stoffel, S, Scharf, S.J., Higuchi, R., Horn, G.T., Mullis, K.B., Erlich, H.A. 1988. Primer-directed enzymatic amplification of DNA with a thermostable DNA polymerase. *Science*, 239, 487-491.

43. Saiki, R.K., Scharf, S., Faloona, F., Mullis, K.B., Horn, G.T., Erlich, H.A., Arnheim, N. 1985. Enzymatic amplification of ß-globin genomic sequences and restriction site analysis for diagnosis of sickle cell anemia. *Science*, 230, 1350-1354

44. Sambrook, J., Fritsch, E.F., Maniatis, T. 1989. Molecular cloning. *A laboratory manual, 2.Edition, CSH*

45. Shibata, D., Martin, W.J., Appleman, M.D., Causey, D.M., Leedom, J.M., Arnheim, N. 1988. Detection of cytomegalovirus DNA in peripheral blood of patients infected with human immunodeficiency virus. *J Infect Dis.*, 158, 1185-1192

46. Sommer, R., Tautz, D. 1989. Minimal homology requirements for PCR primers. *Nucl Acids Res.*, 16, 6749

47. Steffen. C. (ed), 1999. PCR Applications Manual, 2nd edition. Roche Diagnostics GmbH, Mannheim, Germany.

48. Taylor, G.R. 1991. Polymerase chain reaction: basic principles and automation. In PCR A Practical Approach; Eds. Mcpherson M.J., Quirke P., Taylor G.R., *IRL Press*

49. Thein, S.L., Wallace, R.B. 1986. The use of synthetic oligonucleotides as specific hybridisation probes in the diagnosis of genetic disorders. In Human genetic diseases: a practical approach. Eds. Davis K.E. *IRL Press, Herndon, Virginia*

50. Tindall, K.R., Kunkel, T.A. 1988. Fidelity of DNA synthesis by *the Thermus aquaticus* DNA polymerase. *Biochemistry*, 27, 6008-6013

51. Walsh, P. 1992. Preferential PCR Amplification of Alleles: Mechanisms and Solutions. *PCR Methods and Applications*, 1 (4), 241-250

52. Yap, E.P.H., McGee, O'D. 1991. Short PCR product yields improved by lower denaturation temperatures. *Nucl Acids Res.*, 19, 1713

53. White, B.A. (Ed), 1993. PCR Protocols. In Methods in Molecular Biology, volume 15. Hamana Press, New Jersey, USA

Chapter 3

CHAPTER THREE
AMPLICON DETECTION, ANALYSIS AND PCR EVALUATION

1. Introduction

After a PCR assay (or during an assay, with some cycler systems), several crucial questions need to be answered in order to determine the success of the assay. How do we know if the PCR has worked? How should we interpret the results? What do we want/need to do with the PCR amplicons?

There are various ways in which the outcomes of a PCR assay may be determined and it is the nature of the specific PCR that will determine the methods to be used in the evaluation of the assay. These methods are vastly different. Some allow for reading the PCR during the actual run of cycles **(homogenous detection-*see* Chapter 6, Quantitative PCR).** Others can only be used after the amplification cycles have been completed **(heterogeneous detection - this Chapter**).

All the methods have in common the measurement of the integrity and yield of the PCR amplicon, or the rate of accumulation thereof **(Figure 1).** The simplest of these methods entails an electrophoresis assay, in which the amplicon may be separated and visualised on a matrix that allows the separation of nucleic acid molecules. Of these matrices, it is agarose gels that are most commonly used.

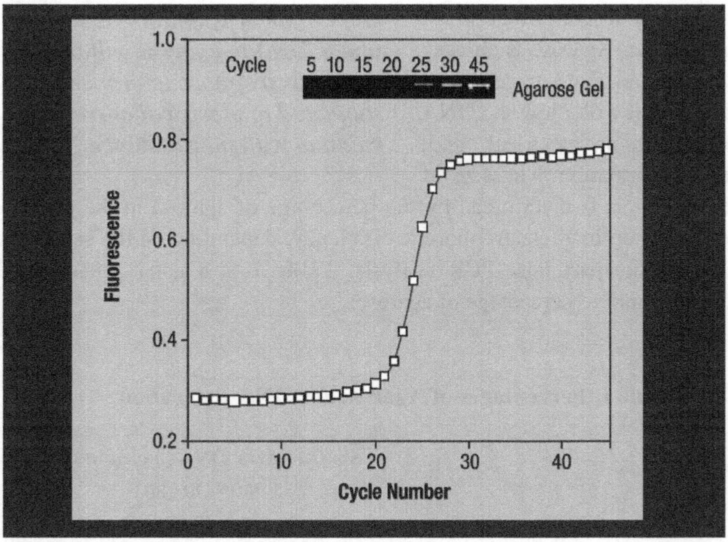

Figure 1. Exponential accumulation of PCR amplicons to reach levels easily detectable by gel electrophoresis and/or fluorescence.

We need to think about the nature of the PCR reaction to understand real time and fluorescence detection of amplicon. The amount of DNA theoretically doubles with every cycle of PCR. After each cycle, the amount of DNA is twice it was before, so after two cycles one has 2x2, as much After 3 cycles, 2x2x2, or 2^3 times as much, after 4 cycles 2x2x2x2 or 2^4 times, as much. Thus, after n cycles, we will have 2^n times as much. But the reaction cannot go on forever and it tails-off and reaches a plateau phase, as shown in **Figure 1.** If we plot these figures in the standard fashion, we cannot detect the amplification in the earlier cycles because the changes do not show up on this scale of detection. Eventually you see a logarithmic or linear phase as they rise above baseline and then a non-linear or plateau phase. Practically, this starts somewhat earlier than we have shown here. If development is plotted on a logarithmic scale, small differences can be observed after earlier cycles. In real time PCR, we use both types of graphs to examine data. Note that there is a straight-line relationship between the amount of DNA and cycle number when you look on a logarithmic scale - because PCR amplification is a logarithmic reaction.

2. Agarose gel electrophoresis (protocol also in Appendix B)

DNA is negatively charged and these molecules will migrate to the positive electrode (cathode) when placed in an electrical field. The ratio of charge to mass in DNA is uniform and therefore all linear DNA fragments will possess the same electrophoretic properties when subjected to electrical current in a medium with no resistance.

Non-linear DNA molecules, like covalently closed circular molecules will have electrophoretic migration patterns that are not directly proportional to mass, due to differences in density and conformation of these coiled DNA species. Agarose is an easy and cost-effective medium that, in gel form, provides the viscosity that will allow the DNA molecules to separate in an electrical field, based on size and conformation. The size of the matrices in agarose gels is determined by the percentage of agarose in the gel and may be manipulated to allow for optimal electrophoretic efficiency, depending on the size ranges of the expected amplicons of the particular PCR analysis. **Table 1** is a good general guideline to determine the most appropriate percentage of agarose:

Table 1. Percentages of Agarose for DNA separation

% Agarose	Size of linear DNA (basepairs)
0.3	1 000 – 30 000
0.7	800 – 12 000
1.0	500 – 10 000
1.2	400 – 7 000
1.3	300 – 3 000
1.4	250 – 3 000
1.5	200 – 3 000
2.0	50 – 2 000

Polyacrylamide gel-electrophoresis (PAGE) can also be used for the separation of DNA molecules. The PAGE system is typically a vertical set-up and the matrix formed by the polymers in the gel is much finer than that of agarose. It therefore offers superior separation of DNA fragments. However, to set up PAGE is much more cumbersome and the electrophoresis time consuming in comparison to agarose electrophoresis. For these reasons, PAGE typically find very limited application in PCR methods and is most used in applications where separation of DNA fragments that differ by as little as a single nucleotide has to be achieved, as for example in DNA sequencing.

2.1 Buffers

Two types of buffers are generally required:

2.1.1. Loading buffer

This is the buffer to be added to the sample of the PCR reaction that will be electrophoresed. This buffer contains a colourant like bromophenol blue that will allow you to visualise the sample when loading the gel and to track the migrating material in the gel during electrophoresis.

2.1.2. Electrophoresis buffer

This is the buffer covering the gel, cathode and anode and contains appropriate ionic salts that will dissociate and travel across the agarose matrix with the movement of electrons from the cathode to the anode.

2.2. Visualising the DNA

After the electrophoresis has been completed there are different methods that may be used to make the separated DNA species in the gel visible to the human eye. The most common of these is to stain the gel with ethidium bromide and then visualize the gel on a UV trans-illuminator. The ethidium bromide intercalates with the DNA molecules and this agent will transmit UV light as longer wave fluorescence that is easy to see and photograph.

2.2.1. Ethidium bromide

Many researchers opt to include ethidium bromide in the loading buffer and/or in the electrophoresis buffer, saving time on the staining step after electrophoresis. This is perfectly acceptable but, because ethidium bromide is an intercalating agent, two important issues should be kept in mind.

Firstly, when ethidium bromide intercalates and associate with the basepairs in the two DNA strands, these helical molecules are rendered stiff and rigid, affecting their migration through the agarose matrix. This effect is not serious, but staining should rather be done after electrophoresis in cases where the accuracy of the size measurement of DNA amplicons is a premise.
Secondly, ethidium bromide is a mutagen.

The intercalated substance will also cause breaks or nicks in DNA strands when exposed to UV (including sunlight). The compound is therefore hazardous and should be handled with extreme care. Other agents for staining include silver, which is very effective, but an expensive alternative.

2.3. Electrophoretic conditions

The conditions of electrophoresis depend on:

- The strength of current (Amperes)
- The power applied (Watts)
- The potential difference (Volts)

These are dependent on the specific application, size of the gel and the percentage of agarose in the gel (density of the matrix).

Most commonly the criterion of 5Volts/cm of gel is applied and results in a total voltage of 60-100 for most gels. In general higher voltages result in faster separation, but lower voltages permit most accurate separation and resolution of different DNA species.

2.3.1. Inspection of gels

After electrophoresis and staining of the gel, inspection on a UV trans-illuminator will allow one to determine:

1. Whether there is an amplification product(s)
2. The quality of the product(s)
3. The specific size of the product(s)

The latter is achieved by the routine inclusion of a DNA standard, that typically contains a range of fragments of known sizes. It is most important however, that the presence of an amplicon after PCR cannot confirm the specificity of the reaction, i.e., the genetic makeup of the product. At the most, the presence of an amplicon of exactly the expected size is an indication that the PCR has targeted the specific template you have intended for it.

Confirmation of the integrity/identity of the amplicon may be achieved by a variety of additional methods, including nested PCR, hybridisation, restriction length polymorphism (RFLP) and sequence analysis.

3. Interpretation of PCR results

All PCR tests are performed together with suitable controls to assist in determining the reliability of the test results obtained.

3.1. A positive PCR test result

This indicated that the nucleic acid genome of the pathogen was present in the sample analysed. A positive PCR result does not necessarily indicate that the agent was present in a viable, proliferating, structurally whole or active form at the time of sample collection.

3.2. A negative PCR test result

This indicates that:

The nucleic acid genome of the pathogen could not be detected under the test conditions used and suggests that the pathogen was either absent at the time of sample collection, or absent in the test material submitted, or present in sub-detectable quantities only.

A negative test result should therefore not be taken as a guarantee of pathogen absence within the animal (or patient) sampled, in a diagnostic application.

A repeat submission of appropriate sample material may, therefore, be advisable.
Note that when the recommended conditions of sample storage and submission are not complied with, a negative effect (reduced sensitivity) on the outcome of the test may occur.

4. Confirming the identity of the PCR amplicon

4.1. Nested PCR

The nested PCR is essentially a PCR in which the template is the amplicon(s) obtained in a previous PCR. However, the reaction is designed as such that an amplicon is sought from two primers that should anneal internally, i.e., within the previously obtained amplicon. If the new set of internal primers do yield the desired product, it is a further indication of the sequence composition (and therefore identity) of the original amplicon.

4.2. Hybridisation: the Southern blot

Hybridisation is carried out after the electrophoresis process with agarose or polyacrylamide gels. The separated DNA fragments are blotted onto nylon or nitrocellulose membranes and fixed to the membrane matrix. This DNA is now available for hybridisation to an isotope or related labelled probe, if it has sufficient homology to that probe. To be effective, a nucleic acid hybridisation probe must have a high degree of specificity.

In other words the probe must hybridise exclusively to the selected target nucleic acid sequence.

Specific probes that will determine the sequence and nature of the amplicon may for example include allele specific oligonucleotides for the human DQ alpha gene, that allows distinction of eight different alleles and twenty-one different genotypes.

4.3. Restriction fragment length polymorphisms (RFLP)

4.3.1. Restriction endonucleases: General information

Restriction endonucleases (RE) are enzymes that cleave DNA in a sequence-dependent manner (protocol: Appendix D).

RE that are generally found in prokaryotic organisms, are probably important for degrading foreign DNA (particularly bacteriophage DNA). Organisms that produce restriction endonucleases protect their own genomes by the methylation of nucleotides within the endonuclease recognition sequences.

There are two major types of restriction enzymes:

Type I enzymes that recognise a specific sequence but make cuts elsewhere

Type II enzymes that make cuts only within the recognition site

Type II enzymes are the most important class and we confine our attention to them. The recognition sequences are generally, but not always, 4 to 6 nucleotides in length and are usually characterised by dyad symmetry (the 5'-3' nucleotide sequence of one DNA strand is identical to that of the complementary strand).

All RE of this class make two single-strand breaks, one break in each strand. There are two distinct arrangements of these breaks:

1. Both breaks at the center of symmetry (generating flush or blunt ends), or
2. Breaks that are symmetrically placed around the line of symmetry (generating cohesive ends)

An important point about these recognition enzymes is this: since a particular enzyme recognizes a unique sequence, the number of cuts made in the DNA from an organism is limited. Because of the specificity, a particular restriction enzyme generates a unique family of fragments from a particular DNA molecule. Another enzyme will generate a different family of fragments from the same DNA molecule.

The family of fragments generated is generally detected by agarose gel electrophoresis of the digested DNA. The fragments migrate at a rate that is a function of molecular weight and their molecular weights can be determined through reference to fragments of known molecular weight that are run concurrently.

4.3.2. Critical parameters

4.3.2.1. Purity of DNA

The efficiency of the RE reaction is very dependent upon the purity of DNA. Contaminants found in some preparations of DNA (e.g., protein, phenol, chloroform, ethanol, EDTA, SDS, high salt concentration) may inhibit RE activity.

The decreased reaction efficiency associated with impure DNA preparations may be overcome by increasing the number of enzyme units added to the reaction mixture (up to 10 to 20 units per μg DNA), increasing the reaction volume to dilute potential inhibitors, or increasing the duration of incubation.

4.3.2.2. RE buffer

For each restriction endonuclease, optimal reaction conditions are recommended by the manufacturer.

4.3.2.3. Other factors

Larger amounts of some enzymes are necessary to cleave supercoiled plasmid or viral DNA as compared to the amount needed to cleave linear DNA.

4.4. Sequence analysis

Compared to all the other methods that really use shortcuts to estimate sequence identity or polymorphism, nucleic acid sequencing and analysis is most definitive, reliable and accurate in determining the exact nature of the PCR amplicon in question. Unfortunately, development costs, running costs and levels of skills required for these methods are high and although automation is possible, the process is at present relatively slow. However, once a sequence database has been generated for isolates of a given pathogen, the sequences of new amplicons can be compared to each other as well as previously published sequences or databases to indicate the type of isolate present in the sample. This technique is therefore hugely successful in studies of the epidemiology of infectious diseases.

There are two general procedures to determine the nucleotide sequence of a gene or genome.

1. The chemical cleavage method devised by Allan Maxam and Walter Gilbert starts with a DNA that is labeled at one end of one strand with ^{32}P. The labeled DNA is then broken preferentially at one end of the four nucleotides. The conditions are chosen so that an average of one break is made per chain and the resulting fragments are analysed on an acrylamide separation gel to be visualised via the radioactive marker on an autoradiogram. This methods found limited application and is hardly ever used.
2. A more popular DNA sequencing procedure is the controlled interruption of enzymatic replication, a method developed by Fred Sanger and his associates. In this method, a DNA polymerase is used to copy a single-stranded molecule (template). This synthesis reaction is primed by a complementary fragment (a sequencing primer, which is very similar to a PCR primer). In addition to the four deoxyribonucleoside triphosphates (which are labeled with radioactivity or with fluorescence markers), the incubation mixture contains a 2', 3'-dideoxy analog of one of the dNTPs. The incorporation of this analog blocks further synthesis of the new chain because it lacks the 3'-hydroxyl terminus needed to form the next phosphodiester bond and therefore terminates the new chain. Hence, different length fragments are generated and analysed on acrylamide gels. This application has seen some tremendous developments. The dideoxy sequencing method has been automated to support the vast load of sequences to be analysed (e.g., human genome project), and the separation system has been modified to allow for several options (e.g., a plate format vs. a gel tube format vs. capillary columns).

5. References

1. Gibbs, R.A., Nguyen, P.N., McBride, L.J., Koepf, S.M., Caskey, C.T. 1989 Identification of mutations leading to the Lesch-Nyhan syndrome by automated direct DNA sequencing of in vitro amplified cDNA. *Proc Natl Acad Sci USA,* 86, 1919-1923

2. Gibbs, R.A., Ngyen, P.N., Caskey, C.T. 1989. Detection of single DNA base differences by competitive oligonucleotide priming. *Nucl Acids Res.,* 17, 2437-2448

3. Greenberg, S.J., Ehrlich, G.D., Abbot, M.A., Hurwitz, B.J., Waldman, T.A., Poiesz, B.J. 1989. Detection of sequences homologous to human retroviral DNA in multiple sclerosis by gene amplification. *Proc Natl Acad Sci USA*, 86, 2878-2882

4. Gyllensten, U.B., Erlich, H.A. 1988. Generation of single stranded DNA by the polymerase chain reaction and its application to direct sequencing of the HLA-DQA locus. *Proc Natl Acad Sci USA*, 85, 7652-7657

5. Mazars, G.R., Moyret, C., Jeanteur, P., Theillet, C.G. 1991. Direct sequencing by thermal asymmetric PCR. *Nucl Acids Res.*, 19, 4783

6. Trower, M.K., D. Burt, I.J., Purvis, et al. 1995. Fluorecent dyeprimer cycle sequencing using unpurified PCR products as templates : development of a protocol amenable to highthroughput DNA Sequencing. *Nucl Acids Res*, 23, 2348-2349

7. Rao. V.B. 1994. Direct Sequencing of Polymerase Chain Reaction-Amplified DNA. *Anal Biochem,,* 216, 1-14

8. Wilson, R.K., Chen, C., Hood, L. 1990. Optimization of asymmetric polymerase chain reaction for rapid fluorescent DNA sequencing. *BioTechniques*, 8, 184-189

Chapter

4

CHAPTER FOUR
LIMITATIONS AND TROUBLESHOOTING

1. General PCR problems

As stressed in Chapter 2, the optimal reaction conditions need to be ensured for each and every PCR application. The reaction conditions will vary greatly, depending on the application, but care should be taken in the use of appropriate primers and primer concentrations, reaction conditions (temperature profile), Mg^{2+} *concentration and the grade of polymerase.*

The following general factors may play a role in problems with PCR or with application of PCR results:

- Thermal cyclers are not equally effective and temperature inconsistencies in the blocks of these machines can be a major detrimental factor in PCR-based assays (Appendix A)

- PCR reactions are very sensitive to reaction conditions and even duplicates might not give quantitatively identical results

- Due to the sensitivity and specificity of the PCR reactions, minor contamination can lead to false-positive signals and false-negative amplifications are often seen

- It has been suggested that templates with a high % G + C content are discriminated against due to low efficiency of strand separation during the denaturation step of the PCR reaction and sequences that are more abundant than others may be preferentially amplified

- Certain PCR target sequences may also be discriminated slightly against others due to selective priming or higher order structure elements that could result in huge differences after multiple cycles. In addition, being an enzymatic process to which exogenous DNA is added, inhibition of the reaction (enzyme) by compounds present in the sample (DNA source) is a serious possibility

1.1. PCR artifacts

Primer-dimer artifacts are frequent in all kinds of PCR reactions. They may be reduced by:

- Avoiding primers that self-anneal or anneal to other primers in the same PCR reaction
- By limiting cycle numbers
- Minimizing cycle lengths
- Increasing annealing temperatures and
- Decreasing concentrations of Mg^{2+}, *Taq* polymerase and primers

1. Primer-dimers occur when 5' to 3' complementarity exists between regions located near the 3' ends of a primer pair. Subsequent annealings between these sites lead to eventual amplification of the remaining ends of each primer. The size of the resultant amplicon will thus be the sum of each primer minus the number of the shared 5'-3' complementary nucleotides. It will contain sequences derived from both primers encompassing the shared region.

2. Primer-dimers may have an additional adverse effect in the case of inverse PCR, viz. they may lead to amplification of the region between the 5'-ends of the primers. Note that the 3'-end strand of a primer-dimer binds to its complement on the template DNA, its 3'extension product copies the sequence lying 5' of the primer to which it binds

3. Serious consideration of these issues is necessary in the primer design process. **(Refer to Chapter 2, Section 2.3.2. and Table 6)**

1.2. Polymerase chain reaction inhibition

Given the multitude of applications for PCR, a great variation of sample types would be expected, especially if investigations involve food, environmental and clinical screening such as in veterinary diagnostics. Common inhibitors in these types of samples include, among others:

- Organic and phenolic compounds,
- Heavy metals
- Lipids
- Certain cations
- Hemoglobin
- Urea

It is of importance to consider the general implications of PCR inhibitors and their elimination.

The mechanisms of PCR inhibition can be divided into three broad categories, based on the point of action within the reaction. These are:

- Failure of lysis or incomplete lysis (where whole cells are used directly as source of template DNA)
- Degradation of the template nucleic acid
- Polymerase inhibition

One of the most common endogenous inhibitors of the polymerase in PCR is insufficiently purified template DNA. It has also been shown that PCR tubes from various manufacturers may give markedly different results, with some of these being strongly inhibitory to the polymerase activity. The powder from gloves and nylon and nitrocellulose filters, commonly used in molecular techniques, can be inhibitory as could be mineral oil that is used to overlay PCR reaction in cases where thermocyclers are not equipped with heated lids.

As mentioned, numerous compounds from clinical, food and environmental specimens can be inhibitory to the polymerase enzyme, although a large number of methods are available to ensure proper removal of such agents during template preparation for the various types of PCR applications.

Several protocols and kits are widely available for the purification of DNA for its use in PCR applications. It is generally recommended that DNA should be purified and should be of the highest quality and integrity possible. Such measure would help to ensure the outcome of the PCR assay and should eliminate the possibilities of reaction inhibition.

1.3. Contamination of PCR

The exquisite sensitivity of PCR often results in problems with reagent or sample contamination. This occurs when exogenous previously amplified sequences, positive control plasmids or DNA from the skin surface of the PCR operator contaminate the reaction mixture resulting in a false-positive signal.

Procedures that aim to enhance the sensitivity of PCR may potentiate this problem. Under normal conditions the greatest risk of contamination is in fact the possibility of introducing a small amount of previously amplified DNA into the new reaction. If, for example, some amplified sequences were carried over in a pipette or found their way into stock solutions they could act as the target DNA for that new reaction. Being specific to the primers, even a tiny contaminant of this nature could make a major contribution to the product of the new reaction. To avoid this possibility, a wise precaution is to set up new reactions using pipettes that have never been used to manipulate post-amplification products. Great care must be taken to avoid storing or manipulating post-amplification products close to PCR reagents.

Reaction constituents can also be sources of contamination and commercial preparations of Taq polymerase have been shown to be contaminated with bacterial DNA. Airborne contamination, such as hair roots or sloughed-off skin cells, are less of a problem in most cases. Although it is obviously best to avoid this possibility, such precautions are often not practical. A minute contribution of foreign genomic DNA to the reaction is unlikely to have a marked effect. However, if the amplification reaction is using a single cell as the source of target then even a single-cell contaminant could be critical.

When undertaking such protocols special precautions must be taken. PCR primers are more resistant to ultraviolet irradiation than target DNA and it has been reported that pre-exposure of a PCR mixture to UV light, for example from a germicidal lamp, can very efficiently eliminate contamination. It may also be possible to remove contamination by pre-treating the PCR mixture with a restriction enzyme or DNase treatment that should cut the PCR product, although these procedures would not remove single-stranded contaminants. In cases where PCR product contamination has become an acute problem, it can be dealt with by designing new primers that flank the original primer sites on the target sequence and hence will not amplify the original product. Of course, this would not address the problem of the source of contamination directly, and other steps will need to be taken to prevent its recurrence.

1.3.1. Avoiding PCR contamination

A number of routine precautions should be taken when performing any type of PCR to eliminate the unwanted transfer of DNA and thus contaminate the PCR:

a) Ideally, experiments should be set up in a laminar flow hood dedicated solely for PCR use. This should be situated in a separate laboratory completely isolated from PCR product stores or plasmid clone preparation areas
b) Separate supplies of pipettors, pipette tips, PCR tubes, and reagents should be kept specifically for PCR and not be used for any other preparations
c) Reagents and equipment such as PCR buffer, distilled water, pipette tips, and PCR tubes should be autoclaved. In general, good laboratory practice and scrupulous attention to technique is required for any PCR-based strategy
d) Oligonucleotide primers should be aliquoted once synthesised. Due to the cost of replacement, special care should be taken with primers; the use of a separate box of pipette tips when aliquoting them is advisable
e) Disposable gloves should be worn at all times and changed frequently, especially when entering or leaving the PCR laboratory, handling concentrated DNA samples or when a splash occurs
f) Powder-free gloves are generally preferred for PCR, since the powder used in powdered gloves may be inhibitory to the PCR reaction and may act as vehicle for DNA contamination
g) Sample DNA or RNA should be added to the reaction last, and it is a good idea to change gloves in between each sample. The spread of contamination picked up by the PCR operator may also be avoided by the wearing of protective clothing that covers the head and face
h) One or more negative reagent controls containing distilled water instead of template should always be included when performing PCR so that carelessness in preparation can be highlighted
i) An extraction control is valuable when dealing with paraffin-embedded material. This comprises a negative control tissue block that is extracted alongside the test samples, thus controlling for contamination prior to setting up the PCR

j) Positive control DNA plasmids should be diluted considerably and added in a different laboratory. Only a very small volume would suffice to contaminate the PCR preparation area and very little is needed to produce a strong positive control signal
k) Nested PCR can also be adopted as an anti-contamination strategy. In addition to its enhanced sensitivity and specificity this method avoids false-positive signals arising as a result of `jumping PCR'. This is where partially overlapping fragments of DNA present in the reaction mixture may extend along each other to form an amplifiable product of the correct length leading to a false-positive result

1.4. Useful pointers to remember

1.4.1. Control experiments

Never do a PCR without proper controls–these include:

- Negative control samples – obligatory
  - Two controls are needed: a water and a negative material control
  - DNA standard in pathogen/target-free test medium (e.g., blood serum) to determine inhibitory properties

- Positive controls – to verify missing, ineffective or degraded components – obligatory
  - Strong positive control – to determine repeatability
  - Weak positive control – to determine repeatability and level of sensitivity

- Internal controls - Amplify two different genes or parts of genes, or genes present in control specimen (cellular gene). These controls are usually in the same tube as your sample

1.4.2. Sources of contamination

1.4.2.1. Before PCR

- Remember: Sterile is not DNA free!
- Human skin, hair etc., contain exogenous DNA that may be contaminants
- Reagents should be PCR grade
- The enzyme preparation can be contaminated
- PCR carry-over contamination (previous PCR's) presents the biggest contamination disaster

1.4.2.2. After PCR

- Amplicons are a most important source of after-PCR contamination therefore; avoid all contact between post-PCR products and new PCR input reagents. Often this is not possible (e.g., nested PCR) but great care should be taken with these procedures
- Do not store amplicons in the freezer. After gel analysis, discard these products
- Robotic systems are increasingly popular due to the limiting of handling and minimisation of cross-contamination offered by such systems

1.5. Prevention and elimination of contamination

- Use an appropriately designed laboratory and follow laboratory rules (Chapter 6)
- Preferably different rooms and operators for pre and post PCR
- Separate autoclave for PCR.
- Laminar flow with UV (UV degrades DNA)
- Use flaming for metal and glass apparatus
- Different operators should have own pipettes and sets of reagents
- Gloves should be changed regularly
- All PCR reagents should be aliquoted and stored at -20°C

1.5.1. Decontamination

UV radiation: The following conditions should effectively destroy exogenous DNA:

- >10 min with 100uJ/min UV (254nm) for tubes and reagents (not the enzyme)
- >8 h with 400uW/cm^2 UV (distance <1m) for decontamination of work surfaces

1.5.2. Chemical/Enzymatic degradation

- Surface decontamination of the workplace and contact areas with 10% Sodium hypochlorite
- Restriction enzyme digestion treatment of reagents.
- Uracil n-glycosylase (UNG) hydrolysis of uracil in amplicons

1.5.2.1. Chemical methods and materials that combat PCR contamination

Most PCR applications can be performed with minimal infrastructure and should be trouble-free if good protocol and common sense is continuously applied. This is particularly achievable if the amount of PCR reactions is relatively limited.

Some laboratories however, need to perform a great many reactions simultaneously. Often such high-throughput laboratories are involved with disease diagnostics where the presence or absence of a given amplicon after the PCR, would be indicative of the presence or absence of a given pathogen.

Clearly, false positives (as result of contamination) and false negatives (inhibition etc.) should be avoided at all costs in such applications. Some specific systems are available to assist in the avoidance of false-positive or false-negative PCR results.

1.5.2.2. Ultraviolet-mediated DNA crosslinking

In this procedure, the PCR tube with all the components for the reaction, except the template DNA, is exposed to UV radiation. This short wave UV radiation (254 and 300 nm together or just 254 nm) then nick and crosslink any contaminating sequences, rendering them un-amplifiable. Typically, such an UV DNA cross-linking method for the destruction of exogenous template could proceed with the following steps.

1. Set up a PCR mix, but do not add sample DNA
2. Transfer PCR tubes to a UV crosslinker (for example, Hoefer Scientific Instruments, model UVC 1000)
3. Set the desired duration of exposure and follow all manufacturers' safety instructions for operating a short-wave UV source. The length of exposure depends on the quantity of likely contamination which in turn depends on the presence of concentrated target DNA, e.g. plasmids, in the laboratory
4. Expose the PCR tubes with open lids
5. Remove from crosslinker, add sample template, seal with liquid paraffin, transfer to thermal cycler, and commence PCR

The disadvantage with this procedure is that the primers and *Taq* polymerase are adversely affected by long exposure to UV, resulting in a subsequent loss of sensitivity to the reaction. If primers can be ruled out as a source of contamination then they need not be exposed for long periods. Following 80 min of UV radiation exposure, up to 25 ng of plasmid DNA cannot be detected by 40 cycles of PCR amplification. Shorter exposure times will deal with lesser amounts of contamination. Higher intensity UV sources will require less exposure.

Routine exposure of PCR buffer, dNTPs, and distilled water to short-wave UV radiation is recommended as an anti-contamination precaution. Exposure of *Taq* polymerase and primers for excessive periods of time is recommended only as a last resort, since this may compromise the sensitivity of the assay.

1.5.2.3. Amplicon Carry-Over Prevention

AmpErase Uracil N-glycosylase (UNG) is a 26 kDA ultra-pure, recombinant enzyme encoded by the *E. coli* uracil N-glycosylase gene that has been inserted into an *E. coli* host to direct the expression of the native form of the enzyme.

When dUTP replaces dTTP in PCR amplification, UNG treatment can prevent the subsequent reamplification of dU-containing PCR products, UNG acts on single- and double-stranded dU-containing DNA by hydrolysis of uracilglycosidic bonds (base excision) at dU-containing DNA sites, releasing uracil and creating an alkali-sensitive apyrimidinic site in the DNA. This activity can result in enhanced PCR specificity. The enzyme has no activity on RNA or dT-containing DNA. UNG and dUTP may also be used to avoid carry-over contamination in RNA PCR. Uracil N-glycosylase can be used to cleave DNA at any position where a deoxyuridine triphosphate has been incorporated.

1.5.3. Use of "Mimic" As Internal Standard to Avoid False Negatives in PCR

To assure the affectivity and fidelity of PCR, a common practice is to amplify the test specimens in parallel with positive internal controls. These usually contain a low number of target molecules and their amplification indicates that all the reagents and the cycling profile was optimal. i.e., the sensitivity of the test was satisfactory. It is a common observation that certain specimens may contain inhibitory ingredients.

Operational errors may also occur during sample preparation or reaction set-up stages. These factors may result in occasional false negative results. A further important aspect is to eliminate the importance played by single copy contaminants by using an internal control, the so-called "mimic". The "mimic" DNA will be considered as the threshold level DNA of the reaction. In short: this would mean that you will have your target DNA and it's two specific primers as well as the mimic DNA and its specific primer pair all in one tube.

During amplicon analysis:

- If you have two amplicons – the sample tests positive for the pathogen
- If you have only the mimic amplicon–the sample tests negative for the pathogen, but the PCR reaction was functional (no PCR inhibition)
- If there is no amplicon – PCR reaction inhibition has occurred

The mimic DNA can be even more useful than the mere fact of just being there to confirm a PCR reaction. For example, mimic molecules that serve as internal controls of amplification and serve as internal sensitivity or quantification controls can be designed.

Mimics have the same primer-binding nucleotide sequences or properties as the real target DNA, but these flank a heterologous sequence of a different size as the target.

The difference in size allows easy discrimination between the PCR products of the mimic and of the real target DNA on agarose gels. Thus, if the two targets are co-amplified, the amplification efficiency in each reaction tube is verified.

To this effect several mimic molecules have been constructed. It has been shown that co-amplification of 5 to 20 mimic molecules with the true targets provide effective internal controls in diagnostic PCR.

2. PCR Troubleshooting

As stressed in previous sections, it is important that PCR applications should be carefully optimised, with regard to all the constituents of the reaction. Naturally, the specific purpose of the assay is also of importance when consideration is given to such test conditions. When the amplification product is to be cloned for example, the highest degree of fidelity is sought. For diagnostic purposes on the other hand, priority is given to a high degree of sensitivity. With this in mind, the following pointers are useful in analysing problems experienced with a PCR assay.

2.1. Misincorporation

Misincorporation is most often promoted under the following sub optimal conditions:

- The total dNTP concentration is very low (<1 μM)
- The concentrations of the four dNTP's are not in balance
- The total dNTP concentration is too high (>200 μM)

2.2. Background

Non-specific amplification products, often present as DNA products of variable lengths (presenting smears on agarose gels), can be a common problem in many PCR's. This background typically results from:

- Too much primer
- Too much enzyme
- Annealing temperature not stringent enough

2.3. Low product yield

A good yield of specific amplicon is crucial to most PCR applications and simplifies detection and characterisation of the product. Low yields of amplicon often results from:

- An incorrect annealing temperature (too high)
- Extension time that is too short
- Enzyme concentration is severely limiting
- Mg^{2+} concentration is not optimal for the assay

2.4 Primer-dimer formation

Primer-dimers, the sum of the two primers (usually 40–60 bp in size) is one unexpected background PCR product that may be obtained **(refer to Chapter 4, Section 1.1).** These non-specific products will decrease the production of desired product and typically results from:

- DNA template left out of reaction
- Too many amplification cycles
- Primers have partial complementarity at the 3' ends

2.5. No product, reaction failure

If absolutely no product is seen, any one or combination of a large variety of factors may be responsible, some of which are decidedly more common than others and listed as such:

2.5.1. Operator failure

Any number of mistakes can be made – often through inexperience and may include:

- Failure to add a reaction component (often the enzyme since this is added last)
- Miscalculation or miss-measurement
- Failure to add oil (where appropriate)
- Incorrect PCR programme, or failure to start the cycler properly

2.5.2. Template issues

Where crude extracts are used, the template can often be responsible for PCR failure:

- Template contains inhibitors
- Template poorly quantified and not diluted correctly
- Template degraded
- Template not present in the extract
- The template and primers are mismatched

2.5.3. Temperature errors

In this regard the annealing and denaturation temperatures are most important:

- A very high denaturation temperature will inactivate the enzyme
- Too low actual denaturation temperature will cause non-denaturation of the template
- Annealing temperatures that are too high will cause PCR failure
- The PCR-cycler is faulty and register incorrect temperatures

2.5.4. Primers

- Primers are degraded. This is however unlikely where good primer management is practiced (Chapter 2).
- Problems with the primer synthesis, e.g., primers are not of the correct sequence or quality, may be impure or may be diluted incorrectly.

2.5.5. dNTP's

The dNTP's should be aliquoted and freeze-thaw cycles limited.

2.5.6. Enzyme

It is unlikely that commercial enzyme preparations should be defective, since the enzyme is stable and good quality control generally applies to commercial supplies.

3. References

1. Akane, C.E. 1994. Identification of the heme compound co-purified with deoxyribonucleic acid (DNA) from blood strains, a major inhibitor of PCR. *J. of Forensic Science*, 39, 362-372

2. Bell, D.A. 1991. Excessive cycling converts PCR products to random length higher molecular weight fragments. *Nucl Acids Res*, 19 (18), 5079

3. Chou, O. 1992. Prevention of pre-PCR mis-priming and primer dimerization improves low-copy-number amplifications. *Nucl Acids Res*, 20 (7), 1717-1723

4. Dieffenbach, C. 1993. Setting up a PCR laboratory. *PCR Methods and Applications,* 3, S2-S7

5. Dragon, E. 1993. Handling Reagents in the PCR Laboratory. *PCR Methods and Applications,* 3, S8-S9

6. Dohner, D. 1995. Inhibition of PCR by mineral oil exposed to UV irradiation for prolonged periods. *BioTechniques*, 18 (6), 964-967

7. Hanni, C. 1995. Isopropanol precipitation removes PCR inhibitors from ancient bone extracts. *Nucl Acids Res*, 23 (5), 881-882

8. Hartley, J. 1993. Dealing with Contamination: Enzymatic Control of Carryover Contamination in PCR. *PCR Methods and Applications,* 3 (2), S10-S14

9. Huang, S. 1994. Long-range jumping of incompletely extended polymerase chain fragments generates unexpected products. *BioTechniques*, 16 (2), 242-246

10. Lee, O.C. 1995. Improved direct PCR screen for bacterial colonies: wooden toothpicks inhibit PCR amplification. *BioTechniques*, 18 (2), 225-226

11. Longo, P. 1990. Use of uracil DNA glycosylase to control carry-over contamination in polymerase chain reactions. *Gene*, 9, 125-128

12. Menon, R. 1991. RT-PCR artifacts from processed pseudogenes. *PCR Methods and Applications*, 1, 70-71

13. Prince, S.E. 1992. PCR: How to Kill Unwanted DNA. *BioTechniques,* 12, 358-360

14. Odelberg, S.J., Weiss, R.B., Hata, A., et al. 1995. Template switching during DNA synthesis by *Thermus aquaticus* DNA polymerase I. *Nucl Acids Res*, 23, 2049-2057

Chapter

5

CHAPTER FIVE
SPECIFIC TYPES OF PCR ASSAYS

1. **Reverse Transcribed PCR (RT-PCR)**
2. **Touchdown PCR**
3. **Nested, Hemi-nested and Pit-stop PCR**
4. **Solid phase PCR**
5. **Multiplex PCR**
6. **Quantitative PCR**

1. Reverse Transcribed PCR (RT-PCR)

In many applications of PCR, the starting template material is not DNA, but indeed RNA. Examples include PCR assays designed for the diagnosis of viral infections where the virus in question has a RNA genome, i.e., HIV, FMDV, rabies and many others. Since RNA cannot serve as a template for PCR, reverse transcription is combined with PCR to make RNA into a complementary DNA (cDNA) suitable for PCR. The combination of techniques is colloquially referred to as RT-PCR.

An example of a RT-PCR protocol is given in **Appendix F**. In some cases PCR's are specifically designed to amplify mRNA, usually to generate a copy of the coding region of a gene by avoiding the noncoding regions (introns) that separate the coding regions (exons) in the DNA of eukaryotic organisms. Furthermore, in targeting RNA with a PCR approach, gene expression may be studied or genes about which very little is known may be cloned (RACE, anchored PCR – see later sections). However, as already indicated, RNA is not a substrate for the Taq DNA polymerases commonly utilised in PCR. To be able to amplify RNA species by PCR, this RNA has to be converted into its DNA complement, or cDNA (complementary DNA). This conversion can be carried out in a simple enzymatic reaction with the enzyme reverse transcriptase (RT) and the DNA product then amplified by PCR.

1.1. Reverse transcriptase

This enzyme is a RNA dependent DNA polymerase able to perform the following reactions:

- Synthesizes a DNA strand on an RNA template
- Removes the RNA strand from the DNA: RNA duplex (RNase H activity)
- Synthesizes a second DNA strand on the DNA template

This enzyme is used in the replication cycle of Retroviruses, a group of viruses that converts their RNA genomes into DNA that will then integrate in the genomes of host cells. The enzyme preparations that are commercially available are therefore typically of retroviral origin, the most common of which are avian myeloblastosis virus (AMV) reverse transcriptase and Moloney murine leukemia virus (M-MuLV) reverse transcriptase. Tth polymerase, an enzyme isolated from the bacterium *Thermus thermophilus*, is a heats table DNA polymerase **(see Tables 8 and 9 in Chapter 2)** with RT activity and can be used in PCR.

Table 1: Properties of Reverse Transcriptases

Enzyme	Maximum size of template	Temperature optimum	Units needed for standard reaction	Ion requirement	Rnase H activity	Use as combined reverse transcriptase and DNA polymerase
AMV Reverse Transcriptase	up to 12 kb	42°C (up to 60°C)	40	Mg^{2+}	++	–
M-MuLV Reverse Transcriptase	up to 10 kb	37°C	40	Mg^{2+}	+	–
Tth DNA Polymerase	up to 1 kb	60°-70°C	4	Mn^{2+}	–	+

1.2. Critical factors in RT-PCR

Great care is needed when working with RNA! RNA is much less stable than DNA and is quickly degraded by RNase exonucleases. These enzymes are ubiquitous and are also heat stable and thus very difficult to get rid of. The method used for the isolation and the protection of the RNA in question should be designed with this in mind. As a general rule all water and buffers used in this process are treated with di-ethyl pyrocarbonate (DEPC) in a process that will destroy RNase. Gloves are to be worn at all times when working with RNA and pipettes, consumables and workspace should be dedicated to RNA work.

1.3. Initiation of the RT reaction - Primers

1.3.1. Oligo $(dT)_{12\text{-}18}$

This primer will anneal to the endogenous poli $(A)^+$ tail on the 3' end of eukaryotic mRNA. Such a primer is used in cases where the PCR is used to randomly clone full-length cDNA of an organism.

1.3.2. Hexanucleotide $d(N)_6$

These primers will anneal to mRNA at any complementary site. Hexanucleotides are commonly used where nothing is known about the RNA template, or in cases where a substantial amount of RNA secondary structure occurs.

1.3.3. Specific oligonucleotide primer

These primers are designed to target the specific initiation point in cases where the sequence to be copied is known.

1.4. The RT reaction

As a general rule, 1 µg of RNA is enough material for a successful RT-PCR. The reverse transcription process and the PCR can be performed in a one-step procedure where all the components for RT and PCR are present from the start, i.e., RNA, primers, dNTP's, PCR buffer, RT and Taq enzymes (or an enzyme that can perform both functions).

Often a two-step procedure is preferred since it offers the opportunity to better control the assay and to perform a reaction titration as needed.

1.5. Examples of RT-PCR applications

RT-PCR has found significant application in the field of disease diagnostics and some of these applications in the field of infectious diseases are discussed in **Chapter 7**. There are other specific types of RT-PCR's that are widely used and one of the most well known, is the anchored PCR.

1.5.1. Anchored PCR

The first cDNA strand is synthesized on a mRNA template after priming with a poly dT primer. The second cDNA strand is then synthesized with a second primer that is gene specific and complementary to the first strand of cDNA. This method is very sensitive and as little as one copy of the specific RNA may be sufficient for amplification. In addition, in cases where the gene sequence is not known, but an internal sequence domain is likely to be conserved (e.g., family specific); an amplicon of this gene may be successfully generated.

2. Touchdown PCR

In touchdown PCR, the PCR temperature cycle is started with an annealing temperature slightly higher than the calculated or anticipated annealing temperature of the primers. A limited number of cycles with these conditions are carried out. Thereafter the annealing temperature is lowered systematically for a limited number of cycles at each temperature profile.

In a typical reaction, each of the PCR temperature cycles will differ by 1°C in the annealing temperature and each of these cycles is run twice. The range of temperatures will typically be over 10-20°C (20-40 cycles) and in the process the touchdown temperature will have been reached and passed.

This basic approach may of course be changed in terms of temperature range, the temperature drop and the individual and total numbers of cycles.

The rationale of this method is that preference is given to the reaction with the highest Tm (and therefore the highest specificity).

For example, if the amplicon with the highest specificity is produced at a Tm 4°C higher than the next Tm specified in the temperature profile, and the Tm is lowered by 1°C every second cycle, then the initial product will have a quantitative advantage of 2^{4x2} (=256) templates.

The method of touchdown has application in cases where the precise Tm of a primer is not known, or when you want to amplify multiplex amplicons using primer sets with different annealing temperatures. This situation often occur in the following cases:

Universal or common primers are used in order to amplify related or cognate sequences from different species or individuals or when attempting the amplification of multiple gene families; where the sequence of the primer is deduced from the amino acid sequence of the protein and the precise nucleotide sequence is thus not known.

The touchdown method is therefore ideal to achieve amplification in cases where a number of mismatches between the primer and the template may occur. Indeed, this method is less sensitive to $MgCl_2$ concentration than the more conventional PCR and is likely to ensure the successful application of a PCR reaction in which conditions are generally sub-optimal. Thus, instead of relying on guesswork or elaborate block titrations, successful amplification may be achieved through this empirical one-step optimization of PCR conditions.

3. Nested, Hemi-nested and Pit-stop PCR

*Nested PCR (nPCR) improves the sensitivity and specificity of DNA (or RNA = nRT-PCR) amplification and may be used to confirm the integrity of a specific amplicon (**see Chapter 4**). Nested PCR means that two pairs of PCR primers are used for a single locus. The first pair amplifies the locus as seen in any PCR experiment. The second pair of primers (nested primers) binds within the first PCR product (**Figure 3**) and produce a second PCR product that will be shorter than the first one (**Figure 4**). The logic behind this strategy is that if the wrong locus were amplified by mistake, the probability is very low that it would also be amplified by the specific internal primers.*

The process utilizes two consecutive PCR's, each usually involving 30 cycles of amplification **(Figure 1).** The first PCR contains an external pair of primers, while the second contains two nested primers that are internal to the first primer pair, or one of the first primers and a single nested primer (so-called hemi-nested primer). The larger fragment produced by the first reaction is used as template for the second PCR where the specificity of the first amplicon is being confirmed by its ability to serve as template for the nPCR **(Figure 1, step 2).** When required, the sensitivity of the nPCR can be enhanced further using 'hot' nPCR. In such application, one of the nested primers is labelled using radio or non-radioactive markers. Typically, the amplicon(s) will be electrophoresed **(Figure. 1, step 3**) and may for example be sequenced to validate identity or for epidemiological investigation **(Figure.1, step 4).**

Pit-stop PCR is a useful modification of nested or hemi-nested PCR, in which two full consecutive cycles are avoided. Instead, the first PCR proceeds for a limited number of cycles (e.g., 10) in order to produce template for a second nested or hemi-nested PCR.

After the first stage, during the "pit-stop", a small volume of the first PCR reaction is withdrawn as template for the second reaction that proceeds for another 20 cycles with the new primer(s) and other constituents. This technique thus offers the speed of a conventional PCR with the sensitivity, specificity and sequence confirmation of a nested PCR.

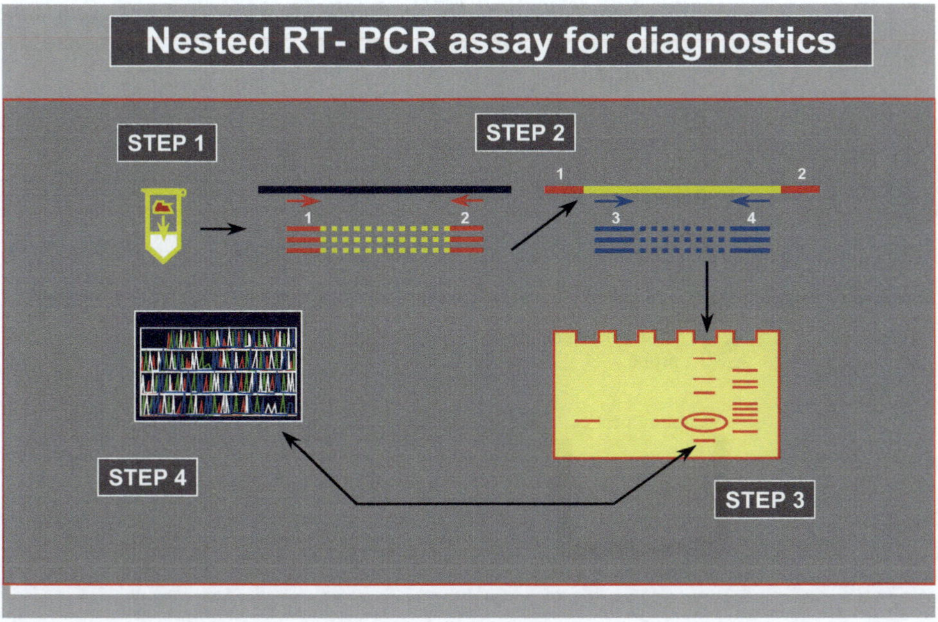

Figure 1. Procedures in a typical Nested PCR.

Step 1: Extraction of target DNA.
Step 2: 1st PCR reaction to amplify the target DNA in a general PCR using the outside PCR primer set (red primers) followed by the 2nd PCR reaction to amplify a fragment of the target DNA in a specific PCR using the inside PCR primer set (blue primers).
Step 3: Analysis of the PCR amplicons by agarose gel electrophoresis.
Step 4: Extraction of the amplicon DNA (band) from the gel, followed by DNA sequencing.

4. Solid phase PCR

As we have discussed, PCR can amplify DNA molecules million fold, but the absence or presence of amplicons must be verified by other methods such as gel electrophoresis, southern blotting, sequencing or various hybridisation techniques. The necessity of characterisation of the amplicons is time-consuming. One technique that offers rapid, sensitive and specific hybridisation, is the enzyme linked oligonucleotide sorbent assay (ELOSA). This is a solid-phase sandwich hybridisation assay that allows a very specific and sensitive detection of a target nucleic acid. As suggested by its name, the protocol of the ELOSA is very close to that of the well-known ELISA and involves only very simple incubations and washing steps. The whole procedure takes about 4 hours

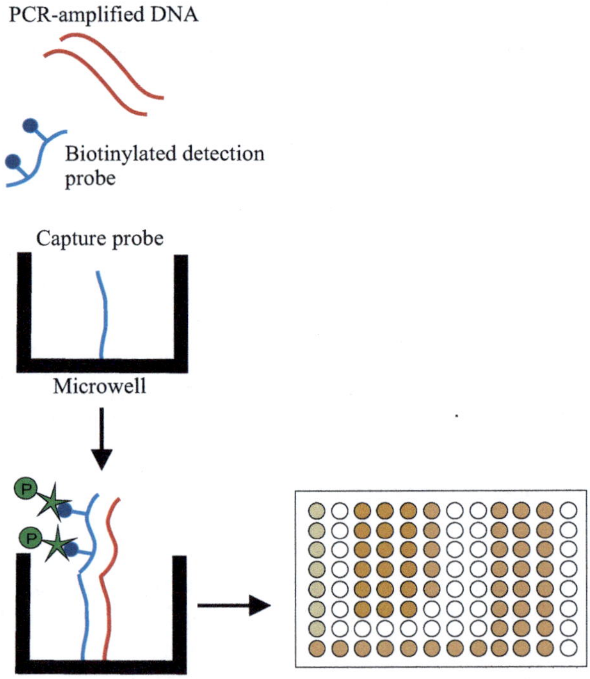

Figure 2. Detection of PCR products using a modified ELISA approach.

1. DNA can be bound covalently to a solid matrix, e.g., CovaLink NH strips. The DNA molecules are bound exclusively at the 5'-end by a phosphoramidate bond. The immobilized primer can serve as a nested primer to increase the PCR specificity
2. Non-radioactive primer probes can be used (downstream primer); biotinylated 5' probes are commonly used. The sensitivity is in the same order as for radioactively labelled probes
3. Hybridisation in microwell plates is, due to easy handling of the microwell plate, a very promising method for diagnostic purposes. Furthermore, microwell plates allow the use of non-radioactive probes, with the further advantage of quantification of the amplicon (Figure 2)

5. Multiplex PCR

*Multiplex PCR can be used to simultaneously screen for different pathogenic strains or isolates, or to differentiate pathogenic and non-pathogenic strains or isolates from each other using target specific primer pairs in combination (**Figure 4**). This gives us the ability to identify a pathogen or a group of pathogens in a population mixture containing other related or non-related organisms. One important requirement is the specific selection of each primer set (to amplify a specific target) and the ability of the diagnostician to identify the pathogen type according to the size of amplicon or according to the sequence.*

5.1. The principle of a multiplex PCR

A diagnostic assay that allows for the simultaneous detection of two or more pathogens is more cost effective and les time consuming than the implementation of multiple single assays for each pathogen.

A multiplex assay may be designed to allow for the detection of various closely related pathogens, where appropriate.

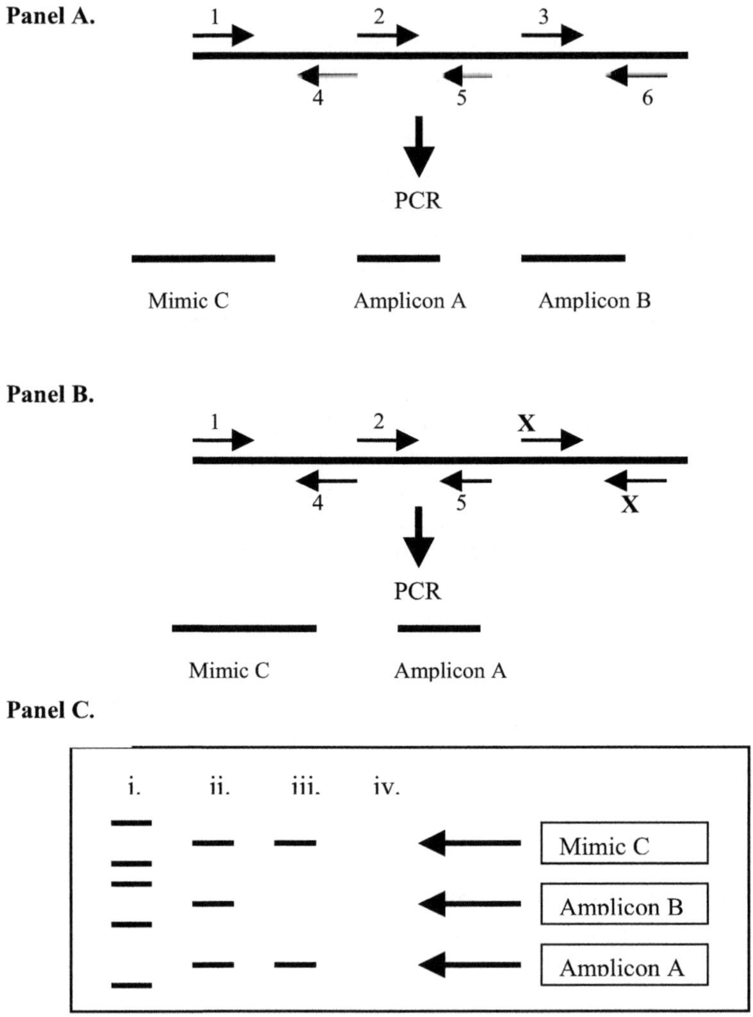

Figure 3. An example of a multiplex PCR assay. The Multiplex differential diagnosis of *Theileria parva.* Mimic C = internal PCR control; Amplicon A = *Theileria spp* (positive control); Amplicon B = *Theileria parva*; **Panel A.** = *Theileria parva* template; **Panel B.** = *Theileria buffeli* template ; Panel C (agarose gel profile)-Lane i. Molecular weight marker; Lane ii. *Theileria parva*, Lane iii. *Theileria buffeli*, Lane iv. Negative control

5.2. Selection criteria

- Select primers with the same Tm for all sites
- Vary length or GC content of reporter probes so that products can be easily separated by Tm

Another important application of the multiplex PCR is known as differential PCR. Differential PCR refers to the amplification of the target gene and a reference fragment in a single reaction vessel. The co-amplification of a quantified reference fragment (added to the sample) and a fragment with an unknown copy number in one reaction tube allows for quantification of the target gene based on a comparison of the amplification levels detected for the reference and the target

6. Quantitative PCR

PCR can amplify a target sequence to levels that are detectable by various traditional techniques, but many researchers will confirm that detection by these methods can be slow and frustrating. A multitude of schemes for the detection of the products of nucleic acid amplification (amplicons) have been introduced since PCR introduction. These detection formats can be divided into two major categories- heterogeneous and homogeneous.

Heterogeneous methods are distinguished by the separation of amplicons from amplification reagents before detection. Homogenous methods allow for the specific detection of amplicons without a separation or wash step. Agarose gel analysis is the heterogenous method that is most generally used ***(see* Chapter 3).**

The general DNA intercalator ethidium bromide (EtBr) is the most commonly used fluorochrome to enable us to visualise the PCR amplicons via ultraviolet light irradiation. This method, though simple and cheep, is time-consuming, labor intensive and prone to carry-over contamination. A homogenous method that can identify the specific amplicon is obviously preferable due to its simplicity, speed and negligible risk of contamination.

Another popular small DNA binding molecule to use is SYBR^R green that binds exclusively in the minor groove of the DNA. Although the binding mechanism of the two dyes is different, they both have increased fluorescence when bound to double-stranded DNA demonstrate no inhibition of PCR.

Different kinds of homogenous detection methods have been designed. Although some chromogenic approaches are in use, most approaches concentrate on a fluorescence basis. Many fluoremetric formats have been described and are in day to day diagnostic use; e.g., those that use fluoremetric probes (so-called hybridisation probes, melting probes or Taqman probes, the use of fluorescent dyes, fluorescence polarisation measurement, combinations of fluorogenic primers, etc.). In this chapter we will discuss quantitative and fluoremetric PCR analysis and its relevance to molecular diagnosis.

During quantitative PCR the DNA segments that are exponentially amplified depend upon the DNA oligonucleotide primers placed into the reaction mixture, and are the basis for specificity of the reaction. Competitor sequences of known concentration are present during the PCR (added to the reactions) and the amounts of analyte product and competitor product are compared in order to obtain an accurate estimation of the analyte's starting concentration.

Quantitation can be semi-quantitative (via comparison with a titration of standards on a gel) or fully quantitative via fluorometric chemistry.

6.1. Traditional quantitative PCR

Accurate quantitation of the target is difficult with normal PCR, as the amount of amplified product does not necessarily reflect the amount of template initially present in the reaction. This is due to the plateau phase (see Chapter 1), that is caused by among others:

- Deactivation of the Taq polymerase
- Shortage of nucleotide substrates
- Shortage of primer
- Inhibition of pyrophosphate
- Re-annealing of amplified DNA's

During the plateau phase of PCR, nearly the same amount of amplified products will be obtained after a sufficient number of PCR cycles, regardless of the initial amount of template. It is almost impossible to determine the PCR conditions that will yield sufficient amount of amplified products for detection by gel electrophoresis before reaching the plateau phase. Consequently, for a group of samples with different template concentrations, quantitation of the target template is difficult.

6.2. Competitive PCR

This technique was developed to overcome the afore-mentioned difficulties in quantitation. In competitive PCR, a known amount of a DNA fragment (competitor) is added to the sample. This competitor must contain sequences for the same primers used to amplify the target.

When the target DNA and competitor are amplified together, both templates will compete for the same set of primers. Because of this competition, the ratio of the amounts of the two amplified products reflects the ratio of the amounts of the target DNA and competitor. Since the initial amount of the competitor is known, the amount of the target DNA can then be estimated according to the T:C ratio.

- T : amount of amplified product from target DNA or RNA
- C : amount of amplified product from competitor

When the T:C ratio = 1, the initial amount of target DNA or RNA will correspond to the amount of competitor **(Figure 4)**

An ideal competitor for quantitative PCR should be:

- A purified DNA template stock of a known concentration obtained at a known quantity or concentration
- Amplified by the same primers as the target DNA.
- Distinguishable from the target DNA (different size, different restriction fragment pattern)

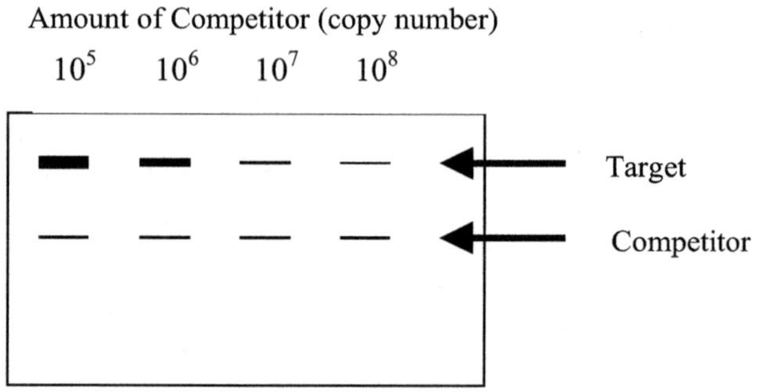

T:C =	10	3.2	1	0.32
Log (T:C) =	1	0.5	0	-0.5

T = amount of amplified product from target DNA or RNA
C = amount of amplified product from competitor

Figure 4. Principles of competitive quantitative PCR.
Samples are analysed by agarose gel electrophoresis and the amount of competitor required to give a T:C ratio of 1 is determined. In this example, the amount of target DNA corresponds to 10^7 copies of competitor.

6.3. Fluorometric real-time quantitative PCR

Fluorometric quantitative PCR analysis (could be analysed after the PCR reaction - use of SYBR^R green chemistry, or during the PCR reaction - real-time PCR using hybridisation and Taqman probes), provides an accurate method for detection of levels of specific DNA and RNA sequences in tissue samples, **(Figures 5-8)**. By plotting the increase in fluorescence versus cycle number, the system produces amplification plots that provide a more complete picture of the PCR process than assaying product accumulation after a fixed number of cycles. It is based on detection of a fluorescent signal produced proportionally during PCR amplification. This chemistry lends itself to detection automatisation and a variety of commercial systems utilise detection robots or testing platforms to analyse amplicon related fluorescence signals. The turn-around time for data acquisition and analysis by on-line real-time PCR is therefore short, in fact almost immediately. Some fluorescence chemistries will be discussed in the following sections. Although fluorescence chemistry was established and well used in automated sequencing platforms, this chemistry with respect to PCR, was initially designed and intended for research purposes. It however became quickly obvious to diagnosticians that the chemistry and platform system had much to offer with respect to turn-around time, repeatability, sample through-put and in limiting contamination.

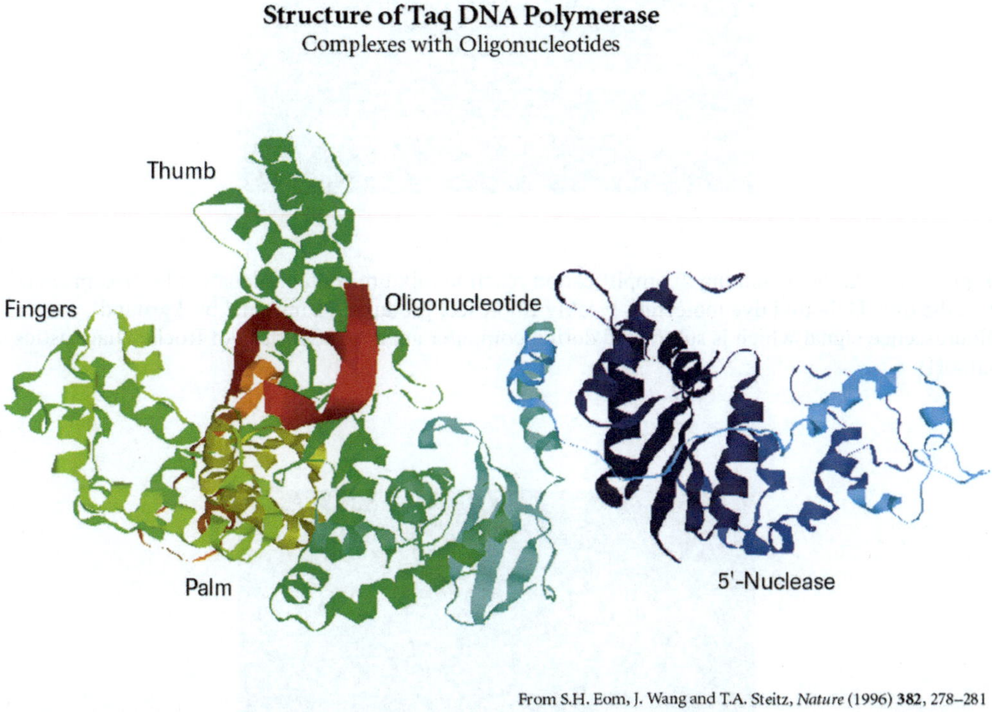
Structure of Taq DNA Polymerase
Complexes with Oligonucleotides
Thumb
Fingers
Oligonucleotide
Palm
5'-Nuclease
From S.H. Eom, J. Wang and T.A. Steitz, Nature (1996) 382, 278–281

Figure 5.

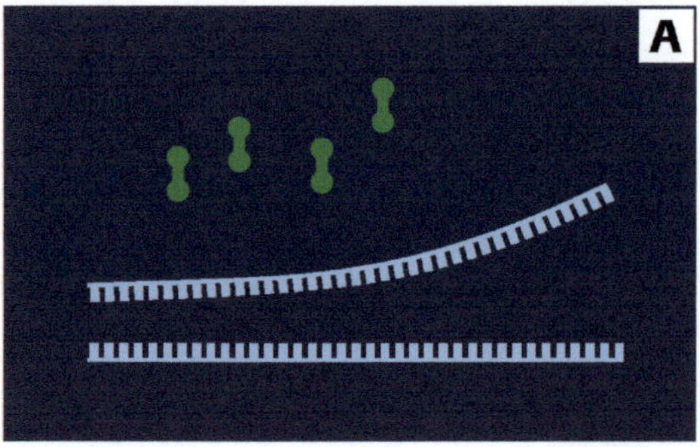

Figure 6A. At the beginning of amplification reaction mixture contains denatured DNA, primers and the dye. Unbound dye molecules weakly fluoresce, producing a minimal background fluorescence signal which is subtracted during computer analysis (Courtesy of Roche Diagnostics, GmbH).

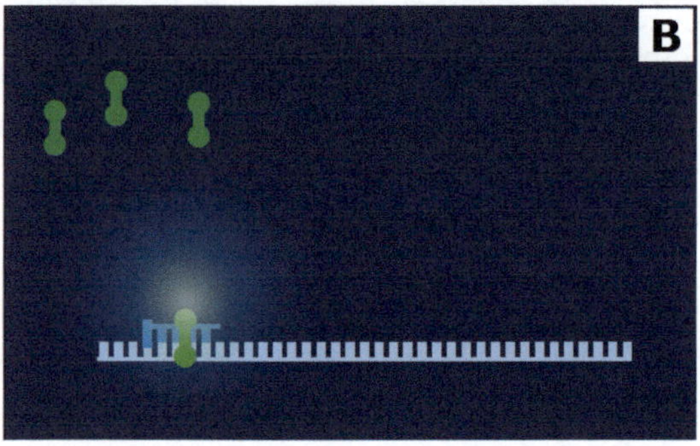

Figure 6B. Fluorescence is greatly enhanced upon DNA-binding. (Courtesy of Roche Diagnostics, GmbH).

Figure 7A and 7B. During elongation, more and more dye molecules bind to the newly synthesized DNA. Upon denaturation of the DNA for the next heating cycle, the dye molecules are released and the fluorescence signal falls. Fluorescence measurement at the end of the elongation step of every PCR cycle is performed to monitor the increasing amount of amplified DNA. (Courtesy of Roche Diagnostics, GmbH).

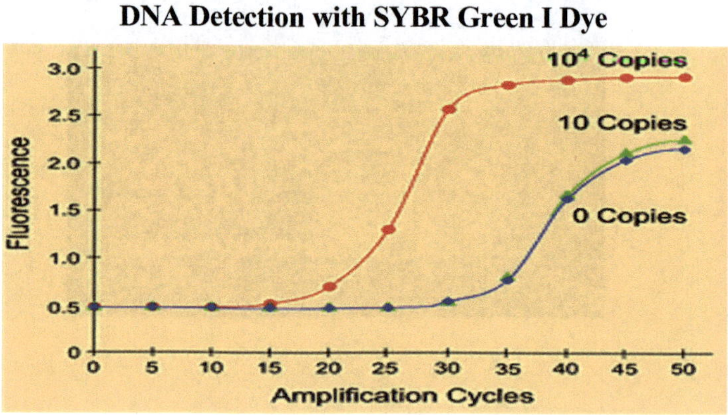

Figure 8. Graphs show fluorescence vs. the number of cycles for differing initial amounts of template copies. When template concentration is high, amplification of specific DNA is observed during early heating cycles. If very few copies of a fragment are to be detected (green graph) unspecific products, usually due to primer dimer formation, can mask the specific signal (lower line on graph). (Courtesy of Roche Diagnostics, GmbH).

6.4. Hybridisation probe chemistry

Real-time systems for PCR were improved by probe-based, rather than intercalator-based PCR amplicon detection. The Roche based hybridisation probes are based on two oligonucleotide probes that bind (1) internally in the PCR amplicon and (2) adjacent to one another. The one probe is labelled at the 3' end with fluorescein while the other probe is 5' end labelled with another fluorochrome. During the annealing phase of the PCR cycle the two probes bind such that an emission signal is detectable, reflecting the quantitative amount of amplicon that was amplified in the previous amplification cycle **(Figures 9-11).** To determine the melting point fingerprint **(Figure 10),** the temperature is increased in a step-wise manner until all emission is decreased. The melting curves can then be determined using special software packages and instrumentation.

6.4.1. Hybridisation probes diagrammatic representation

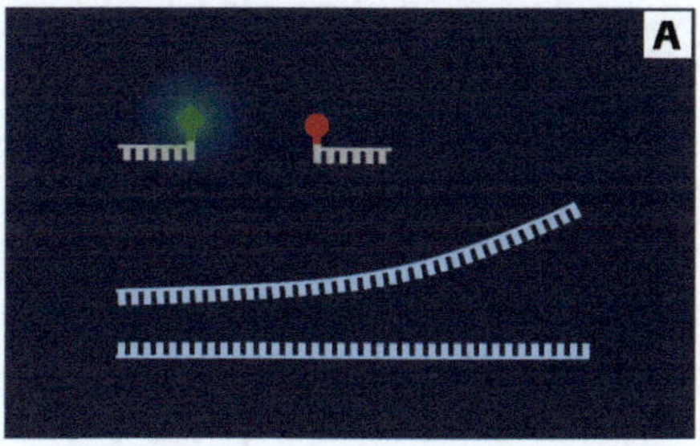

A. The donor-dye probe is labelled with fluorescein at the 3'-end and the acceptor-dye probe is labelled with LightCycler Red at the 5'-end. Hybridisation does not take place during the denaturation phase of PCR and, thus, the distance between the dyes is too larg to allow energy transfer to occur.

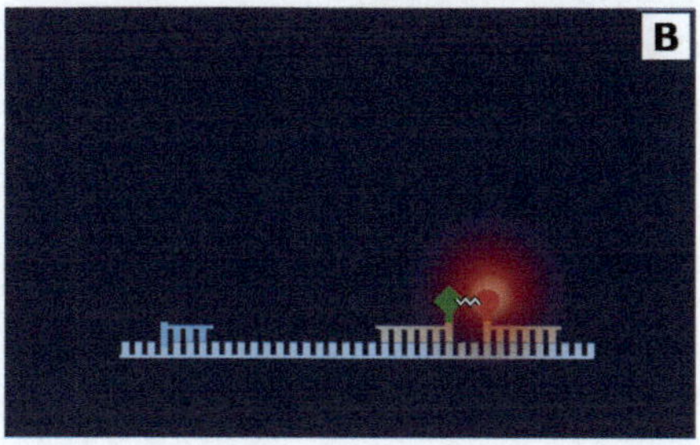

B. During the annealing phase, the probes hybridise to the amplified DNA fragment in a close head-to-tail arrangement. When fluorescein is excited by the light from the LED, it emits green fluorescent light, transferring the energy to LightCycler Red, which then emits red fluorescent light. The red fluorescence is measured at the end of each annealing step, when fluorescence is highest.

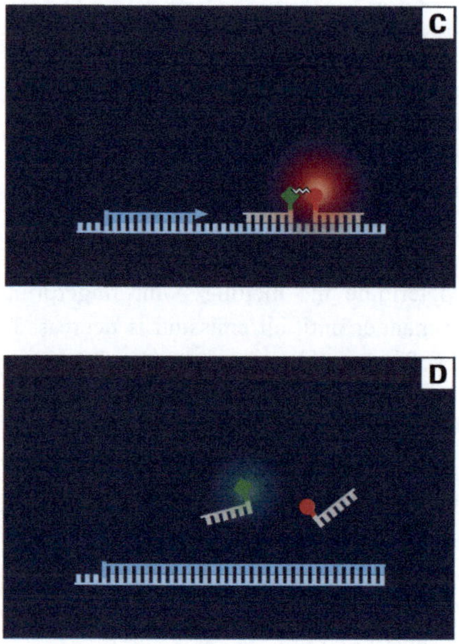

C and **D**. After annealing, the temperature is raised and the HybProbe is displaced during em longation. At the end of this step, the PCR product is double-stranded and the displaced HybProbe probes arew again too far apart to allow FRET to occur.

Figure 9. Hybridisation probe chemistry (Courtesy of Roche Diagnostics GmbH)

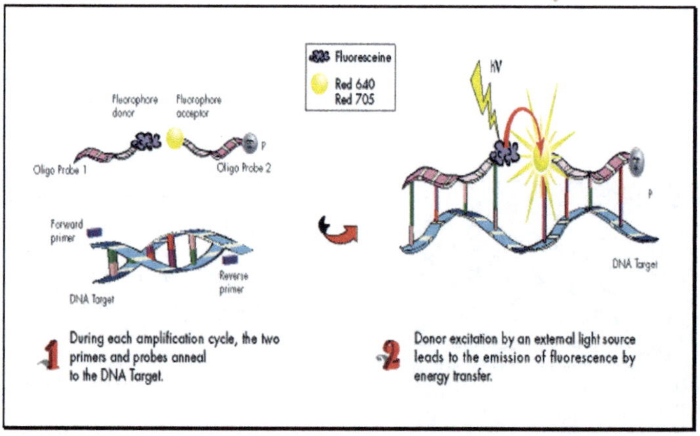

Figure 10. Melting probe chemistry (Courtesy of Roche Diagnostics GmbH)

6.5. Taqman probe chemistry

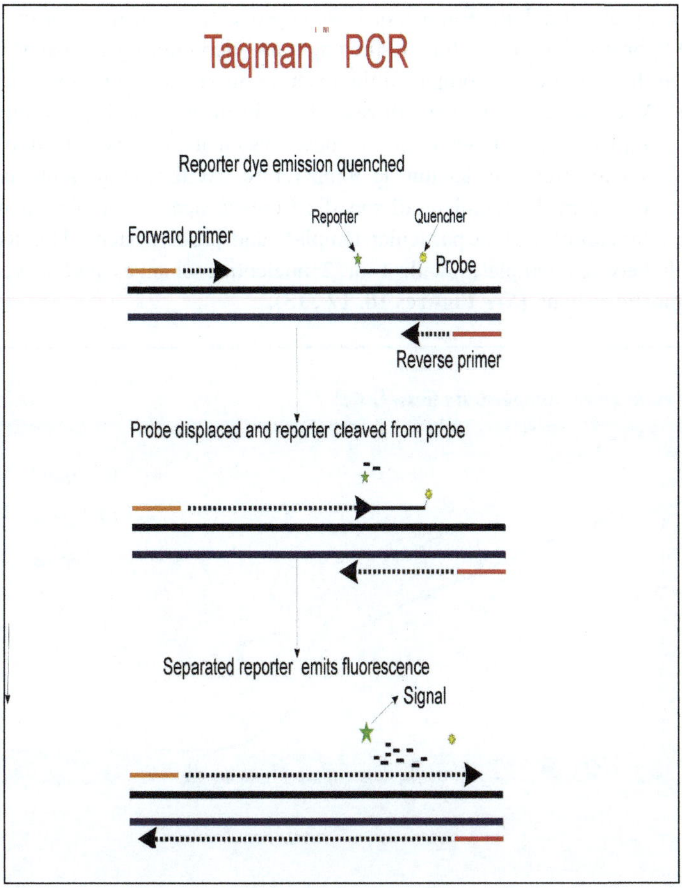

Figure 11. The Taqman probes were designed such that both the fluorochromes (reporter and quencher) are situated on one oligonucleotide probe (Taqman and Beacon probes). This probe is again designed such that it binds internal to the PCR amplicon. During PCR the probe binds to the target template during the annealing step. When the 5' nuclease activity of the Taq polymerase cleaves the probe, the reporter is released, generating a specific emission that can be detected. This chemistry is based on those used by Roche Diagnostics GmbH and Applied Biosystems.

6.6. Melting probe chemistry

Only hybridisation probes (and not Taqman or Beacon probes!) can be used for melting chemistry. In principle, the hybridisation probes bind to the target amplicon during each annealing step of the PCR cycle. When the last cycle is completed the reaction mix is heated up to 94°C to denature all amplicon copies. When annealing is then allowed, the hybridisation probes can bind to one of the single stranded templates. The temperature is now systematically raised from the annealing temperature upwards towards the denaturing temperature. At the temperature point where the probes denature away from the template all specific fluorescence is lost. This temperature is the specific melting temperature of that particular template and probe system. This technique allows us to distinguish between templates with 1 or 2 nucleotide changes and is very useful as a differential diagnostic system. (***See*** **Figures 10, 12 , 13).**

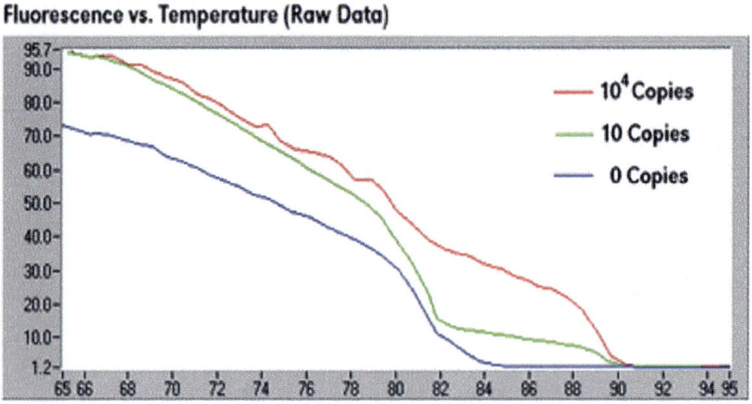

Figure 12. The melting curves of the samples are depicted.
The Y-axis shows the fluorescence and the X-axis shows the temperature. From the different slopes of the plots it can already be concluded that the samples contain different amounts of by-products that melt at a relatively low temperature. (Courtesy of Roche Diagnostics GmbH).

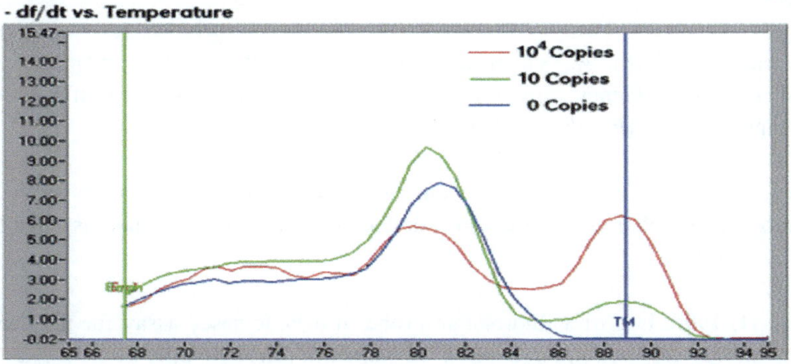

Figure 13. If the first negative derivative (-dF/dT) is plotted against the temperature, the differences between the three samples become more obvious. The peaks on the left-hand side are due to primer dimers that, due to their shorter length, melt at a lower temperature, whereas the peaks at the right-hand side represent the specific amplification product. The peak for the specific product gets smaller when the amount of specific template molecules in the sample is decreased: the sample without specific target molecules (blue curve) does not show any peak on the right side at all. This analysis clearly shows that a specific target sequence exists in the other two samples (green and red curve), as determined by an amplified fragment melting at the expected, characteristic temperature (here approx. 90°C). Melting point analysis is not restricted to enhancing the specificity of DNA quantification with SYBR Green I dye. (Courtesy of Roche Diagnostics GmbH).

7. References

1. Bagnarelli, P., Menzo, A., Valenza, *et al.* 1995. Quantitative molecular monitoring of human immunodeficiency virus type activity during therapy with specific antiretroviral compounds. *J. Clin Microbiol.*, 33, 16-23

2. Wayne, P.A. 1998. Basic cost accounting for clinical services; approved guidelines. November 1998. National Committee for Clinical Laboratory Standards, GP11-A, Vol. 18, Wayne, PA

3. Bassler, H. 1995. Use of a fluorogenic probe in a PCR based assay for the detection of Listeria monocytogenes. *Applied and Environmental Microbiology*, 61 (10), 3724-3728

4. Bisno, A.L., Gerber, M.A., Gwaltney, J.M., Kaplan, E.L., Schwartz, R.H. 1997. Diagnosis and management of group A streptococcal pharyngitis: a practice guideline. *Clin Infect Dis.*, 25, 574-383

5. Brandt, M.E., Padhye, A.A., Mayer, L.W., Holloway, B.P. 1998. Utility of random amplified polymorphic DNA PCR and TaqMan automated detection in molecular identification of Aspergillus fumigatus. *J Clin Microbiol.*, 36, 2057-2062

6. Caetano-Anolles, G., Bassam, B.J., Gresshoff, P.M. 1991. DNA amplification fingerprinting using very short arbitrary oligonucleotide primers. *BioTechnology,* 9, 553-557

7. Carman, W.F., Jacyna, M.R., Hadzizyannis, S., Karayinnis, P., McGravey, M.J., Makris, A., Thomas, H.C. 1989. Mutation precenting formation of hepatitis B e antigen in patients with chronic hepatitis B infection. *Lancet II*, 588-591

8. Carman, W.F., Kidd, A.H. 1989. An assessment of optimal conditions for amplification of HIV cDNA using Thermus aquaticus polymerase. *J Virol Methods,* 23, 277-289

9. Cole, S.A. and J.E. Hixson. 1995. Baboon lipoprotein lipase: cDNA sequence and variable tissue-specific expression of two transcripts. *Gene*, 161, 265-269

10. DiCesare, J. 1993. A High-Sensitivity Electrochemiluminescence-Based Detection System for Automated PCR Product Quantitation. *BioTechniques,* 15 (1), 152-156

11. Edwards, M. 1994. Multiplex PCR: advantages, development, and applications. *PCR Methods and Applications*, 3, S65-S71

12. Eom, S.H., Wang, J., Steitz, T. A. 1996. Structure of Taq DNA polymerase. Nature, 382, 278-281

13. Espy, M.J., Smith, T.F. 2000. Detection of cytomegalovirus (CMV) in clinical specimens by LightCycler PCR (abstract). Abstracts of the 100th General Meeting of the American Society for Microbiology, No. C-62. Los Angeles, CA, May 21-25, 2000

14. Espy, M.J., Patel, R., Paya, C.V., Smith, T.F. 2001. Quantification of Epstein-Barr virus (EBV) viral load in transplant patients by LightCycler PCR (abstract). Abstracts of the 101st General Meeting of the American Society for Microbiology, No. C- 148. Orlando, Florida, May 20-24, 2001

15. Espy, M.J., Rys, P.N., Wold, A.D., Uhl, J.R., Sloan, L.M., Jenkins, G.D., Ilstrup, D.M., Cockerill, F.R. III, Patel, R., Rosenblatt, J.E., Smith, T.F. 2001. Detection of herpes simplex virus DNA in genital and dermal specimens by LightCycler PCR after extraction by IsoQuick, MagNA Pure, and BioRobot 9604. *J Clin Microbiol.*, 39, 2233-2236

16. Ferre, F. 1994. Quantitative PCR: An Overview. The Polymerase Chain Reaction. Ed. Mullis. Boston: Birkhauser

17. Ferre, F. 1992. Quantitative or Semi-quantitative PCR: Reality Versus Myth. *PCR Methods and Applications*, 2 (1), 1-9

18. Frohman. M. 1994. On beyond classic RACE (rapid amplification of cDNA ends). *PCR Methods and Applications*, 4 (1), S40-S58

19. Gibson, U.E.M., Heid, C.A., Williams, P.M. 1996. A novel method for real-time quantitative RT-PCR. *Genome Res.*, 6, 995-1001

20. Haff, L. 1994. Improved quantitative PCR using nested primers. *PCR Methods and Applications*, 3 (6), 332-337

21. Heid, C.A., Stevens, J., Livak, K.J., Williams, P.M. 1996. Real-time quantitative PCR. *Genome Res.*, 6, 986-994

22. Heroux, J. 1995. Quantitative Analysis of Specific mRNA Transcripts Using a Competitive PCR assay with Electrochemiluminiscent Detection. *PCR Methods and Applications*, 4 (6), 327-330

23. Holland, P. 1991. Detection of specific polymerase chain reaction product by utilizing the 5'->3' exonuclease activity of Thermus aquaticus DNA polymerase. *PNAS*, 88, 7276-7280

24. Holland, P.M., Abramson, R.D., Watson, R., Gelfand, D.H. 1991. Detection of specific polymerase chain reaction product by utilizing the 5'-3' exonuclease activity of Thermus aquaticus DNA polymerase. *Proc Natl Acad Sci USA,* 88, 7276-7280

25. Hoof van Huijsduijnen, R.A.M., Ayala, G., DeLamarter, F. 1992. A means to reduce the complexity of oligonucleotides encoding degenerate peptides. *Nucl Acids Res.*, 20, 919

26. Horn, G.T., Richards, B., Klinger, K.W. 1989. Amplification of highly polymorphic VNTR segment by the polymerase chain reaction. *Nucl Acids Res.,* 17, 2140

27. Ishaq, M. 1994. Differential expression of mRNAs for JC virus large and small tumor antigens in brain tissues from progressive multifocal leukoencephalopathy patients with and without AIDS. *PNAS,* 91, 8283-8287

28. Jessen-Eller, K. 1994. Quantitation of metallothionein mRNA by RT-PCR and chemiluminescence. *BioTechniques*, 17 (5), 962-972

29. Jin, C. 1994. Rapid construction of deleted DNA fragments for use as internal standards in competitive PCR. *PCR Methods and Applications*, 3, 252-255

30. Lee, O.C 1993. Allelic discrimination by nick-translation PCR with fluorogenic probes. *Nucleic Acids Research*, 21 (6), 3761-3766

31. Livak, K. 1995. Towards fully automated genome wide polymorphism screening. *Nature Genetics*, 9, 341-342

32. Livak, K. 1995. Olignucleotides with Fluorescent Dyes at Opposite Ends Provide a Quenched Probe System Useful for Detecting PCR Product and Nucleic Acid Hybridisation. *PCR Methods and Applications*, 4 (6), 357-362

33. Livak, K.J., Flood, S.J.A., Marmaro, J., Giusti, W., Deetz K. 1995. Oligonucleotides with fluorescent dyes at opposite ends provide a quenched probe system useful for detecting PCR product and nucleic acid hybridisation. *PCR Methods Appl.*, 4, 357-362

34. Lockey, C., Otto, E., Long, Z. 1998. Real-time fluorescence detection of a single DNA molecule. *Biotechniques.* 24, 744-746

35. Loeffelholz, M.J., Thompson, C.J., Long, K.S., Gilchrist, M.J.R. 1999. Comparison of PCR, culture, and direct fluorescent-antibody testing for detection of *Bordetella pertussis. J Clin Microbiol.*, 37, 2872-2876

36. McGoldrick, A., Lowings, J.P., Ibata, G., Sands, J.J., Belak, S., Paton, D.J. 1998. A novel approach to the detection of classical swine fever virus by RT-PCR with a fluorogenic probe (TaqMan). *J Virol Methods*. 72, 125-135

37. Murray, P.R., Washington, J.A. II. 1975. Microscopic and bacteriologic analysis of expectorated sputum. *Mayo Clin Proc.*, 50, 339-344

38. Murphy, G.F. 1993. Rethenium Complexes as Luminescent Reporters of DNA. Methods in Enzymology. *Anonymous Academic Press*, 226pp. 576-595

39. Myers, R. 1995. Amplification of RNA: High-Temperature Reverse Transcription and DNA Amplification with Thermus Thermophilus DNA Polymerase. *PCR Strategies. Ed. Innis. San Diego: Academic Press. Inc.*, p 58-68

40. Osterrieder, N., P.H. Hubert, C. Brandmuller, et al. 1994. A touchdown PCR for the differentiation of equine herpesvirus type 1 (EHV-1) field strains from the modified live vaccine strain Rach. *J. of Virol Methods*, 50, 129-136

41. Reischl, U., Wittwer, C. & Cockerill, F. (Ed), 2002. Rapid cycle real-time PCR. Methods and applications. Springer-Verlag Berlin Heidelberg, Germany.

42. Roux, K. 1994. Using mismatched Primer-Template Pairs in Touchdown PCR. *BioTechniques*, 16 (5), 812-814

43. Roux, K. 1995. Optimization and troubleshooting in PCR. PCR Methods and Applications. 4 (5), S185-S194

44. Ruster, D.W. 1995. Quantification of Hepatitis C Virus RNA by Competitive Reverse Transcription and Polymerase Chain Reaction Using a Modified Hepatitis C Virus RNA Transcript. *Anal. Biochem.*, 224, 597-600

45. Uhl, J.R., Sandhu, G.S., Kline, B.C., Cockerill, F.R. III. 1996. PCR-RFLP detection of point mutations in the catalase-peroxidase gene *(katG) of Mycobacterium tuberculosis* associated with isoniazid resistance. In: Persing DH (ed) PCR Protocols for emerging infectious diseases. *American Society for Microbiology*, Washington, DC

46. Wilkinson, E. 1995. Development of Competitive PCR and the QPCR System 5000 as a Transcription-based Screen. *PCR Methods and Applications*, 4 (6), 363-367

47. Wilkinson, E.T., Cheifetz, S. and Degrandis S.A. 1995. Development of competitive PCR and the QPCR system 5000 as a transcription-based screen. *PCR Methods and Applications*, 4, 363-367

48. Wu, C.H., Wang, C.C. Yang, H.M. 1995. Sequences at four termini of the giardiavirus double stranded DNA. Gene, 158, 129-131

Chapter

6

CHAPTER 6
THE PCR LABORATORY

1. The laboratory set-up
2. Quality assurance systems for the diagnostic PCR laboratory
3. The use of laboratory accreditation with reference to GLP compliance monitoring

1. The laboratory set-up

The arrangement or design of a PCR laboratory is of great importance in maintaining a high standard of performance. As a minimum you need three dedicated areas. If more space is available, it may be used to good advantage, in properly separating the different processes involved in your PCR assays.

The minimum layout requirements for a PCR laboratory are illustrated in **Figure 1**, whereas the ideal PCR laboratory set-up is shown in **Figure 2**.

The minimum layout arrangements include three dedicated areas which are separate specimen preparation areas and an area dedicated to PCR.

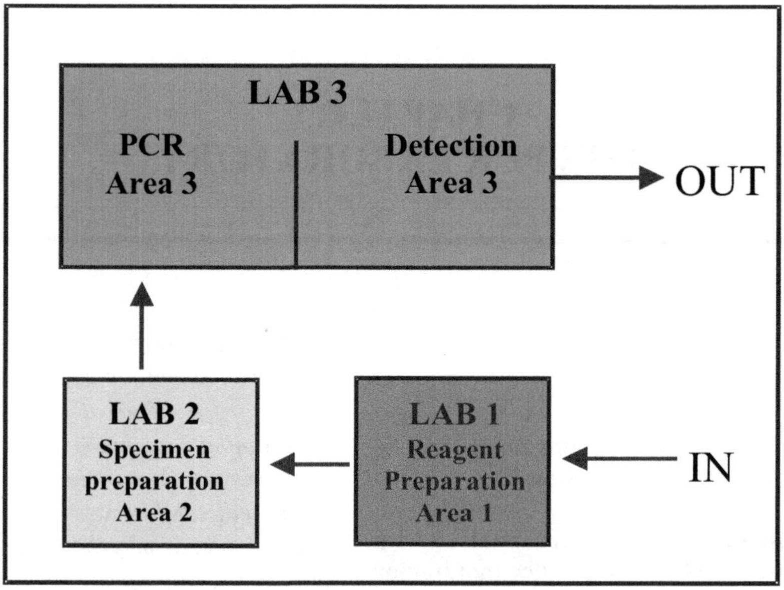

Figure 1. Minimum layout requirements for a basic PCR laboratory

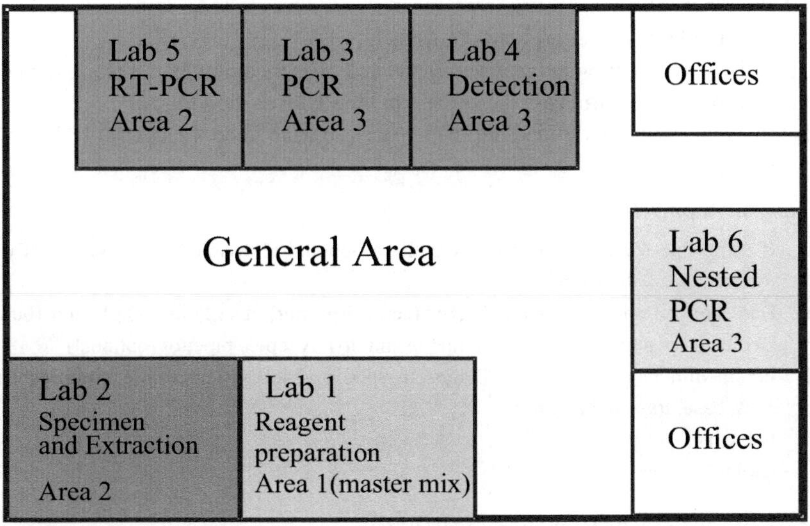

Figure 2. Ideal physical arrangement for a PCR laboratory

1.1. Area 1

1.1.1. Reagent Preparation

The first steps in the PCR procedure starts in Area 1.
Here, master mix (amplification reagents) reagents and primers are prepared along with controls. Prepared master mix and controls are then added to the PCR reaction tubes.
If possible this area should be divided into two spaces so as to keep the control DNA away from the master mixes.

1.1.2. Equipment required in Area 1

a. If using one room, a dead-air box with an optional ultraviolet (UV) light or, if using two rooms a clean, dedicated area
b. Dedicated micropipettes with plugged (aerosol-barrier) tips (25 μL, 50 μL, and 100 μL) or
c. Dedicated positive-displacement pipette and tips. A repeat pipettor (optional)
d. Perkin-Elmer GeneAmp™ PCR System 9600 or 2400, MicroAmp™ consumables (tubes, caps, base, tray, and retainer)
e. Lab coat
f. Gloves

1.1.3. Important considerations applicable to Area 1

With PCR diagnostics, always **work in a one-way direction** from Area 1 to Area 2 to Area 3 in order to avoid carryover contamination from amplified products

- Area 1 should be located as far away from Area 3 as possible
- Area 1 should **not** be in close proximity to the centrifuge area (Area 2) to avoid possible aerosol contamination
- Clean, dedicated area working surface **must be decontaminated prior** to use with 10% bleach solution (freshly made daily) followed by a 70% ethanol rinse or equivalent
- If a UV box is used, for your own safety, you **must turn off** the ultraviolet light in the box *before* placing your arms and hands into the working area
- Master mix and controls are to be prepared in this area
- Prepared master mix and controls will be added to PCR reaction tubes and capped in this area
- Dedicated pipettes with plugged (aerosol-barrier) tips or positive-displacement tips, or a repeat pipettor are required in Area 1 and should **not be transferred** to either Area 2 or Area 3
- Lab coats designated for wear in Areas 1 and 2 should **never** be worn into Area 3 to avoid carryover contamination of amplified product from Area 3 back to Area 1 or Area 2:-two coats may be used, disposable coats may be worn over a cloth coat
- Gloves must be **worn at all times** for your safety as well as for control of contamination from one area to another
- Gloves are to be changed at each of the three work areas
- Gloves worn in Area 3 must **never** be worn in Area 1 or Area 2

1.2. Area 2

Specimens are prepared in this area and are added to PCR tubes with master mix. Specimens are now considered amplification-ready.

1.2.1. Equipment required in Area 2

1. Biological cabinet
2. Standard clinical tabletop centrifuge and/or microcentrifuge (max RCF 16,000g)
3. Dedicated micropipettes with plugged (aerosol-barrier) tips (25 µL, 50 µL, and 100 µL) or dedicated positive-displacement pipette and tips
4. Repeat pipettor (optional)
5. Stopper remover or gauze pads
6. 2-mL screw-cap Sarstedt or equivalent tubes
7. Sterile calibrated transfer pipettes
8. 15-mL conical polypropylene tubes
9. Dry-heat temperature block(s)
10. Vortex mixer
11. Lab coat (same coat as in Area 1 or a different coat)
12. Gloves (new pair)
13. Spectrophotometer

1.2.2. Important considerations applicable to Area 2

There is no need for a dedicated biological cabinet; however, remember:

- Working surface in cabinet **must be decontaminated prior** to specimen preparation with 10% bleach solution (freshly made daily) followed by a 70% ethanol rinse or equivalent
- Cabinet **must** be on 30 min **before** specimen preparation procedure
- All pipettors, pipettes, bulbs, and other equipment used in specimen preparation should be kept in this cabinet at all times
- Specimens must **be stored separately** from reagents-two separate refrigerators are required.
- Dry baths are preferred over water baths water baths can contaminate specimens through seepage into poorly stoppered tubes
- A set of pipettes should be **dedicated exclusively** to be used for specimen preparation
- To avoid specimen-to-specimen contamination, plugged (aerosol-barrier) pipette tips or positive-displacement tips, **must** be used in Area 2
- Lab coats designated for wear in Areas 1 and 2 should **never** be worn into Area 3 to avoid carryover contamination of amplified product from Area 3 back to Area 1 or Area 2.
- Gloves must be **worn at all times** for your safety as well as for control of contamination from one area to another
- Gloves are to be changed at each of the three work areas, gloves worn in Area 3 must **never be worn** in Area 1 or Area 2

1.3. Area 3-Amplification (PCR) and Detection

The final steps in the procedure takes place in Area 3. Here, amplification-ready specimens undergo PCR and the PCR products (amplicons) of this PCR are then detected. This area should also be divided into two areas if possible.

1.3.1. Equipment required in Area 3

a. Perkin-Elmer GeneAmp™ PCR System 9600 Thermal Cycler or equivalent
b. Dedicated micropipettes with plugged (aerosol-barrier) tips (25 μL and 100 μL) or dedicated positive-displacement pipette and tips
c. Multichannel pipettor (optional). Repeat pipettor (optional), Nonplugged pipette tips (optional)
d. Microwell plate lid
e. Disposable reagent reservoirs
f. Key for microwell strip removal (optional)
g. Incubator (37°C ± 2°C) with or without CO_2
h. Microwell plate washer (optional)
i. Microwell plate reader
j. Printer
k. Lab coat (different from the one worn in Areas 1 and 2)
l. Gloves (new pair)
m. Electrophoresis apparatus and documentation system
n. Microwave oven

1.3.2. Important considerations applicable to Area 3

- Area 3 should be kept as far away as possible from Areas 1 and 2 to avoid aerosol contamination
- Incubator temperature should be kept stable
- Traffic in and out of incubator should be reduced to a minimum
- Area 3 pipettes should **never** be used in Areas 1 or 2
- Pipettes with plugged (aerosol-barrier) tips are used for pipetting denaturing solution into PCR tubes and denatured amplified product into microwell plates
- Non-plugged tips may be used for all other reagent additions a pipette contaminated with this highly concentrated product could cause false-positive results plugged tips therefore are required to prevent this potential carryover contamination
- Plate washer is not required; manual washing of plates is acceptable.
- Plate reader should have a printer for a hard-copy readout; a computer interface is optional.
- **As this is a one-way workflow**, the lab coat worn in Area 3 must **never** be worn in Areas 1 and 2
- Gloves must be **worn at all times** for your safety as well as for control of contamination from one area to another
- Gloves are to be changed at each of the three work areas
- Gloves worn in Area 3 must **never** be worn in Area 1 or Area 2

1.4. Special considerations when performing PCR for diagnostic purposes

In addition to the standard biosafety laboratory practices and good laboratory protocol you already perform in your laboratory.

The following considerations apply specifically to PCR diagnostics and should be monitored in order to ensure both your safety and the integrity of your assay.

1. Always work in a one-way direction **from Area 1 to Area 2 to Area 3** to avoid carryover contamination from amplified products
2. Specimens must be stored separately from reagents so as not to contaminate open reagents
3. When handling potentially contaminating material, always use a pipette with a plugged (aerosol-barrier) tip or a positive-displacement tip (refer to Areas 1, 2 and 3 for steps in which such pipettes are required)
4. To avoid possible aerosol contamination, all centrifuges should be kept at a distance from areas where you are preparing master mix and controls and adding prepared specimens to master mix
5. To avoid specimen contamination from seepage into poorly stoppered tubes, water baths are to be avoided; dry baths or blocks, if needed, are preferable
6. Lab coats must be worn in all areas; the coat worn in Areas 1 and 2 may be the same
7. **The coat worn in Area 3 must never be worn in Areas 1 and 2!**
8. Gloves must be worn **at all times** for your safety as well as for control of contamination from one area to another
9. Gloves are to be changed at each of the three work areas
10. Gloves worn in Area 3 must **never** be worn in Area 1 or Area 2

1. 5. The PCR Worksheet

Ref. nr: ....

Test: ....

<u>For:</u> Name: ....
Department ....
Institution: ....
Address ....
....

Tel: ....

Fax: ....

email: ....

Test by: ....

Date received: .... Date processed: ....

Sample history:
....
....

Laboratory history of sample:
....
....

Result: ....

Notes: ....
....

1.6. Checklist of required items

1.6.1. In Area 1

- Dead-air box with an optional ultraviolet (UV) light
- Dedicated micropipettes with plugged (aerosol-barrier) tips or dedicated positive-displacement pipette and tips
- Repeat pipettor (optional)
- PCR consumables: Perkin Elmer 9600
- dNTPs, 10 x buffer +Mg, tubes, caps, base, tray, and retainer
- Lab coat.
- Gloves

1.6.2. In Area 2

- Biological cabinet
- Standard clinical tabletop centrifuge microcentrifuge (max RCF 16,000)
- Dedicated micropipettes with plugged (aerosol-barrier) tips or dedicated positive-displacement pipette and tips
- Repeat pipettor (optional)
- Stopper remover or gauze pads
- 2 mL screw-cap Sarstedt or equivalent tubes
- Sterile calibrated transfer pipettes
- 5 mL conical polypropylene tubes
- Dry-heat temperature block(s)
- Dry bath (37°C) or water bath (optional)
- Vortex mixer
- Lab coat
- Gloves

1.6.3. In Area 3

- Perkin-Elmer GeneAmp™ PCR System 9600 or 2400 Thermal Cycler
- Dedicated micropipettes with plugged (aerosol-barrier) tips dedicated positive-displacement pipette and tips
- Multichannel pipettor (optional)
- Repeat pipettor (optional)
- Nonplugged pipette tips (optional)
- Microwell plate/washer/lid/reader and printer
- Disposable reagent reservoirs
- Key for microwell strip removal (optional)
- Incubator (37°C +/- 2°C) with or without CO_2
- Electrophoresis apparatus
- Documentation system (UV + scanner + computer + printer)
- Lab coat
- Gloves

2. Quality assurance systems for the diagnostic PCR laboratory

The OIE defines validation as the evaluation of a process to determine its fitness for a particular use. A validated assay will then be expected to consistently identify animals as positive or negative for the analyte, process or infectious agent being sought. The validation process of an assay forms part of a quality system whereby veterinary testing laboratories comply with standards (e.g., the OIE or ISO Standard) for management and technical competence. This process will then form the basis of a quality assurance programme and the accreditation of the laboratories performing the test. The implementation of quality systems is increasingly becoming an international necessity. In this regard, accountability to third parties especially with the control of major animal epizootics is of particular importance. With increasing trade and trade agreements, importing countries require assurance that animals and animal products are free of disease causing agents. Validated assays and quality assured procedures have to be used for this purpose.

2.1. Establishment of the PCR assay

The selection of a PCR assay must be based on factors including scientific and international acceptance, cost, available resources, nature of the intended use, sensitivity and specificity, number of tests to be done and availability of standard reagents.

Optimization of the assay must then be performed and should include a series of experimental procedures and the evaluation of the data generated. Analysis should determine a fixed protocol for use, the nature and number of controls required and specifications required of reagents and equipment.

2.2. Validation of the assay

Validation establishes the performance standards of the assay (e.g., sensitivity, specificity, accuracy, precision, positive/negative cut-off criteria etc) using appropriate statistical methods.

Validation can include a comparison with other methods, with reference standards, collaborative studies with other laboratories using the same test procedure, experimental challenge studies or reproduction of data from accepted standard methods or a reputable publication.

2.2.1. Specificity

This is the ability of the assay to exclusively detect the agent of interest. Clinically this means the probability that a healthy animal will be correctly identified as such. In the case of PCR, this means the ability to detect the genome of the relevant pathogen and would also require that genetically similar agents be tested for cross-reactivity.

2.2.2. Sensitivity

This is described as the ability to precisely detect small quantities of the test agent. Clinically this means the ability to correctly identify an animal with the disease and with PCR, the ability to detect pathogen-specific nucleic acid. In the case of screening tests used in surveys, a high sensitivity assay is initially used with positive cases being re-tested with a more specific assay. Of importance would be the ability to detect the agent at levels where critical health-related decisions must be made.

The OIE has formulated a standard procedure for the process of validating a diagnostic laboratory. It has to be accepted that the criteria for serological test validation will be different and that the criteria for nucleic acid validation has not been finalised yet. Many researchers and committees (OIE and WHO) are presently setting-up validation guidelines for PCR. We would like to herewith suggest workable and achievable guidelines that are after-all practical in the PCR laboratory.

2.2.3. Stage 1 validation

Involves the development of the assay performing a feasibility study to determine whether the assay can detect a range of agents (e.g., virus concentrations, virus serotypes/genotypes), without background activity. This stage include all aspects of development and optimisation of the test, including the identification of the test, the determination of the target template(s), the sequence determination of the primers and the test conditions and criteria.

2.2.4. Stage 2 validation

Involves the evaluation of the test against a panel of known positive and negative template samples to determine the analytical sensitivity and specificity.
It also involves further development and standardisation of the assay in order to optimise all its components. Repeatability of results should then be established before continuing with the validation process (stage 3). In addition:

- Analytical sensitivity is needed to determine the smallest amount of virus that can be detected using end-point dilution
- Analytical specificity determines whether there is cross-reactivity with heterologous viruses not targeted for detection
- Internal quality control will involve the continuous monitoring of the assay for assessing repeatability and accuracy

The following stages are performed less frequently or otherwise less completely, although in the case of Stage 3, it is the most important component of the validation process.

2.2.5. Stage 3 validation

This involves the determination of diagnostic sensitivity and specificity as elucidated in the following section.

2.2.6. Diagnostic sensitivity and specificity

Diagnostic sensitivity refers to the proportion of known infected reference animals testing positive [TP/(TP+FN)], while diagnostic specificity refers to the proportion of uninfected reference animals that test negative [TN/(TN+FP)]. (TP = true positive; FN = false negative; TN = true negative; FP = false positive).

The infection status is determined by a relevant golden standard assay. In the case of serological assays, the OIE recommends that 300 reference samples from known infected animals and not less than 1000 reference samples from known uninfected animals be included.. It is even recommended that these values be increased to 1000 and 5000 respectively, to increase accuracy.

In the case of PCR, these values have not been determined but will most likely consist of significantly lower numbers. We propose that the PCR test is validated using field samples and field conditions in parallel to determine the diagnostic sensitivity and specificity.

The number of samples will depend on the type of disease but we propose a minimum of 20 samples for rare pathogens and 100 for more common pathogens.

Repeatability refers to the amount of agreement between replicates within or between runs. Reproducibility compares the same assay as performed between different laboratories. It is recommended that at least 10 samples representing the full range of expected virus concentrations be tested in duplicate (known titres, low through high). The extent of agreement between a test value and the expected value for a sample of known virus concentration will reflect the accuracy of the assay.

2.2.7. Stage 4 validation

This stage entails the continued monitoring of the validity of assay performance in the field by calculating the predictive value of positive or negative results based on estimates of virus prevalence in the target animal population. This can only be done satisfactorily if diagnostic sensitivity and specificity data are available.

2.2.8. Stage 5 validation

Involves the maintenance of validation criteria using internal quality controls. Frequent monitoring for repeatability and accuracy are needed. The OIE also recommends biannual ring-testing to determine reproducibility between laboratories, although annual testing is also described. A validated assay should consistently provide test results that identify animals as being positive or negative and accurately predicts the infection status with a predetermined degree of statistical certainty.

2.3. Establishing a formal quality assurance programme or accreditation

The intention of accreditation or the formal recognition of a quality assurance programme by an independent third party, is to ensure that good laboratory practice, quality control and quality assurance is in place within the diagnostic laboratory. Specific standards are used (e.g. ISO 17025) in accordance with the requirements of the inspecting body (e.g., a National Laboratory Accreditation Service).

Requirements will include a quality policy, a quality manual, procedures for evaluating and reviewing new and existing test methods and protocols, the use of blind quality assurance within and between laboratories, monitoring of equipment, ongoing training, an audit programme, regular review of the quality system and documentation and records of all quality activities and evaluations. Standard operating procedures (SOP's) will be required for not only the test methods used, but also for the calibration and maintenance of instrumentation and equipment used.

2.4. Proficiency testing

Proficiency testing is the means used to determine the capability of a laboratory to perform the assay and effectively detect the agent (internal proficiency testing). Such testing will also contribute to ensuring that within or between laboratories performing routine diagnostic services, a specific assay is performed according to established international standards (external proficiency testing). It is also intended to achieve standardisation of the assays in question. This will ensure that test results obtained are reproducible within and between laboratories and will therefore give a measure of confidence that the results obtained using such an assay are reliable and trustworthy.

2.4.1. Internal proficiency testing

Internal proficiency testing is used to monitor the ability of members within a laboratory to produce repeatable and accurate results.

2.4.2. External proficiency or ring testing

This is used for inter-laboratory comparisons and forms part of Stage 3 and 5 of the validation process as formulated by the OIE. This can be performed on a round-robin (continuous) basis. Although it is a tedious and costly process, it is a necessary activity to achieve standardisation of PCR tests performed. The procedure here will be that a reference laboratory periodically sends each participating laboratory an external proficiency test, a panel of blind coded samples representing the full range of expected concentrations of the pathogen, as well as material derived from an uninfected source. Each participant then processes and tests the samples according to a particular assay method used in-house and (or at a following stage, using an agreed protocol). Statistical comparisons are then made among the laboratories.

The full range of virus that might be encountered in a clinical specimen should be tested. Samples from uninfected sources which test positive in 2 or 3 of the 3 laboratories can be discarded and may suggest mislabelled or contaminated samples. The goals of the clinical programme will also influence the criteria for proficiency.

2.4.3. Precision

The precision of an assay is assessed by comparing results from coded replicates of the same sample.

2.4 4. Reproducibility of results

Reproducibility is established when samples are assayed repetitively. Samples that form part of the coded blind panel can be in duplicate. Yearly or if possible, biannual testing should be performed and should be repeated more frequently by the laboratory that fails the criteria agreed upon. Internal proficiency testing should be performed more frequently, e.g., quarterly. It may be advisable to run a large number of known negative samples on a periodic basis, with a more modest number regularly (negative water, and no RT control for each principal sample). Reports should be written on an annual basis.

A false negative will be assigned to a result if all dilutions from an infected source test negative. A false positive is considered when at least one of all the dilutions from an uninfected source tests positive. In the case of Newcastle disease virus testing, the estimate of sensitivity is based preferably on pfu, but the use of HA units is probably a more practical approach although this only gives an approximation of the amount of viral RNA present. RNA degradation can take place whilst HA titre may not have been significantly affected.

An individual, whose function it will be to prepare and allocate a coded identity and store the reagents on a routine basis, is identified from the coordinating laboratory. In addition, such panels will have to be regularly given to each member on a rotating basis. At least ten samples should be used, consisting of a combination of cDNA, a number of serially diluted positive test samples and if possible, a large panel of negative samples. If results for a specific laboratory remain within acceptable limits, and are accurate and reproducible, it can be certified as an official laboratory for that assay (e.g., by the European Union). Interpretation of data will indicate whether a laboratory or a specific person and the specific assay used are successful on a routine basis. Corrective measures can then be established. This may entail comparisons in methodology between laboratories and between personnel; the introduction of an alternative assay or protocol or reagent; or the retraining of laboratory personnel.

The success of the procedure will depend on the voluntary co-operation of all participants. The emphasis should be on facilitating the learning process for all, with the intention of ensuring an efficient and professional service to the client. Frequent group discussions, creative alternatives and problem solving activities are likely to facilitate the remedying of problems encountered and should be based on an effective team effort where all members are willing to contribute constructively.

2.4.5. Control samples

Control samples should be routinely included with each batch of test samples assayed during the proficiency testing phase of validation. Control specimen samples will have been shown to be positive or negative by at least one classical diagnostic assay according to an internationally accepted golden standard. In addition, several repeat PCR assays should have confirmed the outcome. Ideally, in the case of a positive control sample, it would be particularly useful if the PCR amplicon derived from a positive control sample were sequenced and the resultant sequence data shown to correlate with that derived from the genome of the relevant target agent (assuming sequence data for comparative purposes is available). This, however, is not always possible.

2.4.5.1. Positive control samples

These should include:

- Specimen material with a high pathogen concentration
- Specimen material with a low concentration (preferably the lowest limit routinely detected, or preferably the lowest amount of detectable target agent present in sample material by conventional means and by reported PCR data
- Specimen material with expected pathogen concentration in an infected case (not essential)
- Extracted positive DNA/RNA can be used only if the above are not available

In the case of a proficiency testing process, specimen material spiked with a dilution series of virus (e.g., 10^{-1} to 10^{-10}) should also be included. Virus titre should have been established by means of a golden standard assay.

2.4.5.2. Negative control samples

These should include:

- A negative control consisting of all reagents except extracted sample DNA or RNA-this should confirm absence of amplicon or agent itself or other cross reactive entities in reagents

- A negative specimen control sample (allantoic fluid, buffer, tissue), if available, that will undergo nucleic acid extraction to confirm absence of contamination in the extraction procedure

- Extracted nucleic acid derived from pathogen-free sample material obtained previously (e.g., tissue or other medium normally submitted, including tissue, blood, buffer)-should confirm that no non-specific binding to virus-free specimen DNA/RNA has taken place (eukaryotic and prokaryotic genomic DNA or RNA present in sample). This control is not essential

- In the case of RT-PCR, an additional negative control consisting of all reagents as well as the extracted sample nucleic acid, but *without* reverse transcriptase included, this should confirm absence of DNA or amplicon contaminants in both reagents *and* test material

3. Use of laboratory accreditation with reference to GLP compliance monitoring

Principles of good laboratory practice (GLP) have received substantial attention in recent years and have been discussed in depth by many working groups and international committees. One such leading committee is the chemicals group and management committee of the organisation for economic co-operation and development (OECD). In general; a laboratory must satisfy an appointed and internationally accepted inspection body of its quality of service. This quality service will be based on performances recognised worldwide as the benchmark of excellence.

A laboratory shall establish procedures for monitoring compliance with GLP Principles, based on laboratory inspections (usually directed by the National inspection body) and study audits directed by accreditation criteria (for example EU criteria based on ISO 17025 guidelines). This authority or authorities are mandated to discharge the functions required by the procedures for monitoring compliance and will require that the management of a test facility issues a declaration, where applicable, that a study was carried out in accordance with GLP Principles and pursuant to any other provisions established by national legislation or administrative procedures dealing with good laboratory practice.

The primary task of the national GLP compliance monitoring authority is to verify the validity of such declarations made by laboratory management. GLP compliance monitoring not only examines the procedures and practices used by a test facility to carry out safety studies on chemicals or chemical products, (e.g., industrial chemicals; pharmaceuticals; veterinary drugs; pesticides; food and feed additives; and cosmetics) but also evaluates performance, i.e., the integrity of the data and the reconstructibility of the study. The driving force for GLP compliance monitoring is the requirement to assure regulatory authorities that data they receive in testing studies can be relied upon when making assessments of validity of data and its applicability to disease diagnosis.

3.1. Laboratory accreditation

Laboratory accreditation specifies and underwrites the technical competence of a laboratory to carry out specified determinations to defined standards on a continuing basis.

A laboratory accredited for such determinations (e.g., physical-chemical and analytical procedures) according to ISO/IEC Guide 17025 or equivalent standards can be considered to have satisfied many of the GLP requirements. However, certain fundamental requirements of the

GLP Principles are not covered by laboratory accreditation according to standards and criteria. Other requirements, while called for during laboratory accreditation, are much more stringent under GLP; these are related to recording and reporting of data and management of data retained in archives to allow complete retrospective investigations.

Special emphasis should be placed on key procedures for field studies, such as test item storage, data collection in the field, application equipment calibration, test item application, and specimen collection and transportation. This may require an approved study plan amendment if the method has not been fully developed or validated at the time the original study plan was signed.

4. References

1. Belák, S., Ballagi-Pordány, A. 1993. Application of the polymerase chain reaction (PCR) in veterinary diagnostic virology. *Vet. Res. Commun.*, 17(1), 55-72

2. Belák, S., Ballagi-Pordány, A. 1993. Experiences on the application of the polymerase chain reaction in a diagnostic laboratory. *Mol. Cell. Probes*, 7(3), 241-248

3. Ballagai-Pordány, A., Belák, S. 1996. The use of mimics as internal standards to avoid false negatives in diagnostic PCR. *Mol. Cell. Probes*, 10(3), 159-164

4. Dieffenbach, C. 1993. Setting up a PCR laboratory. *PCR Methods and Applications*, 3, S2-S7

5. Tyagi, S., Kramer, F.R. 1996. Molecular beacons: probes that fluoresce upon hybridisation. *Nature Biotechnol.*, 14(3), 303-308

6. Vet, J.A., Majithia, A.R., Marras, S.A., et el. 1999. Multiplex detection of four pathogenic retroviruses using molecular beacons. *Proc. Natl. Acad. Sci., USA* 96(11), 6394-6399

7. OIE, Manual of Standards for Diagnostic Tests And Vaccines (4th Edition), 2002

8. Van Devanter, D.R., Warrener, P., Bennett, L., et al. 1996. Detection and analysis of diverse herpes viral species by consensus primer PCR. *J. Clin. Mocrobiol.*, 34(7), 1666-1671

Chapter

7

CHAPTER SEVEN
DISEASE DIAGNOSIS USING PCR SPECIFIC PROCEDURES FOR IMPORTANT VETERINARY PATHOGENS

---

1. **Reception of diagnostic samples for PCR assays**
2. **Recording and dissemination of PCR results from diagnostic samples tested**
3. **The detection of Newcastle disease virus using a Triple One-step RT-PCR assay**
4. **The detection of brucella species and strains using a PCR assay**
5. **The detection of Malignant catarrhal fever virus using a PCR assay**
6. **Detection of Foot and mouth disease virus, Rinderpest virus, Peste-des-petits-ruminants virus and Bovine viral diarrhoeal virus using a RT-PCR assay**
7. **The detection of Mycoplasma using a PCR assay**
8. **The detection of Rift valley fever virus using a PCR assay**
9. **The detection of Lumpy skin disease virus using a PCR assay**

1. Reception of diagnostic samples for PCR[2]

Aim

The purpose of these procedures are to ensure that diagnostic samples submitted to the PCR laboratory (e.g., via the Registration Section), are efficiently recorded, distributed and stored, and that those who are involved with performing the PCR test procedures are suitably notified.

1.1. Personnel

Responsible person nominated e.g., secretary, programme manager, other members of the PCR laboratory.

1.2. Authorised users

All PCR laboratory personnel unless otherwise specified.

1.3. Safety

1.3.1. Precaution 1

Samples removed from their packaging are handled with latex gloves. The samples are placed onto the spillage tray on the table designated “Samples received”.

1.3.2. Precaution 2

After labelling sample or sample batch (with a laboratory number as a sequential number according to the record file) using a water-resistant marker pen, it is then placed into a plastic bag. This is then preferably deposited into the -70°C freezer (alternatively, if space is limiting, into a -20°C walk-in freezer).

1.3.3. Precaution 3

No food or drink is placed into the fridge. Hands are to be washed after handling samples

[2] Courtesy of Onderstepoort Veterinary Institute, South Africa

1.4. Accompanying Documents

For example:

- Newcastle disease virus PCR SOP
- Dissemination of results SOP
- Tabulated records, registration forms and photo documents
- List of control samples, validation and proficiency records, signature list, sensitivity and specificity data

1.5. Equipment

- 4°C Fridge and -70°C Freezer
- Photocopier

1.6. Materials and accessories

- Pens, marker pens
- Plastic bags
- Latex gloves

1.7. Methods

Samples received (together with a sample submission or registration form) are registered at the Registration Office and all necessary information recorded.

- A copy of the submission form is archived in the registration office
- A registration number is allocated and noted on the submission form
- The registration number is recorded in a sample book, and the sample and submission form taken by a messenger to the PCR laboratory
- Any sample not distributed within a reasonably short period of time must be stored at 4°C, e.g. In the case of ostrich tracheal swabs submitted for Newcastle disease virus PCR testing, these are to be deposited in a –20°C (if available, a -70ºC) freezer when there is a delay of more than several hours (e.g., overnight or over weekends)
- Freeze-thawing cycles should be avoided
- The sample is then sent by messenger to the laboratory
- The messenger has the responsibility of delivering the sample to the secretary in the Reception Office, or to the Programme Manager.
- The submission form is scrutinised by the receiver and the nature of the assay required noted

Should there be any discrepancy (e.g., the form requests another assay not performed in the division; or wrong sample format), then the sample is returned to the Registration Office.

Once samples have been received at the laboratory, the record book of the messenger is signed by the receiver.

1. The samples are placed into the 4°C fridge in the Reception Office, should there be any delay in processing
2. A record of minimum and maximum fridge temperature is also made daily. Samples must be deposited in a -20°C freezer (or if possible, a –70°C freezer) as soon as possible
3. The submission form is deposited into the "Registration Forms Received" tray
4. At a convenient time, the appropriately labelled file for samples submitted for a particular assay is retrieved from the Reception Office "PCR Record Files" cupboard. The files retrieved contain sample registration forms, photo documents with a worksheet and a summarised record of all samples received. The registration forms are ordered according to their laboratory numbers. A laboratory number is then allotted to the sample
5. The sample is then removed from the refrigerator or freezer, numbered with the laboratory number using a permanent waterproof pen. Latex gloves can be worn during this procedure
6. A temporary record is kept that indicates the date of reception and the receiver, the registration number and laboratory number, and sender/owner names.
7. At a suitable time, preferably daily, sample details are recorded electronically (or on hard copy) in a record file
8. This file is to be regarded as the formal record of all samples received and their results

The details included are:

- Date of reception
- Registration number
- Allocated laboratory number
- Sender/owner names
- Number of samples
- Number of test required

In addition, results will also be recorded here, together with an indication as to where the photo document is kept (it will usually be attached to the registration form with the highest laboratory number indicated in the photo document).

Further comments may include the status of samples received, any request for further submission, reference to further tests etc. Such comments are usually indicated in an abbreviated form (abbreviations indicated in footnotes) because of space constraints.

PCR laboratory personnel who are free to perform an extraction, will then fetch the sample from the refrigerator or freezer.

2. Recording and dissemination PCR results from diagnostic samples tested

Aim.

The purpose of these procedures is to ensure that test results are efficiently and accurately made available to the submitter of the samples and/or the owner of the animal concerned. Time delays should be kept to a minimum. Suitable records are essential for this to occur.

2.1. Personnel

Responsible person: Members of the PCR laboratory.

2.2. Authorised users

All PCR laboratory personnel and Division Secretary.

2.3. Accompanying Documents

For example: Newcastle disease virus PCR SOP

Dissemination of results SOP

2.4. Procedures

A photo record is made of the gel, indicating the outcome of the PCR reaction. This photo must be appropriately marked by the person performing the PCR assay.

The following details are noted:

- Date
- Laboratory number
- Label for each well with relevant details viz. sample number, positive, negative and water control, 100 bp molecular marker

Where there are space constraints, relevant data is recorded on an accompanying work sheet which will also specify the procedure used and other data pertaining to the performance of the assay. The photo-document is placed into the "Results tray" in the Reception Office, or presented to the Programme Manager.

The Programme Manager is responsible for the final analysis of the results.

The relevant test file is retrieved from the PCR Record Files and the sample registration form located.

- Relevant data will be recorded electronically or on hard copy. This data will include:
- Test outcome
- Location of photo-document which is usually behind the registration form with the highest laboratory number as shown on the photo document
- Payment to be requested
- Comments concerning further assays or re-submissions required etc

The latter is usually indicated in abbreviated form (due to space constraints) according to abbreviations indicated in the footnote. If access to the electronic file is not readily available, then data can be recorded temporarily in a table inserted in the front of the file.

Usually this is not necessary since all relevant data can be readily obtained from the documents inserted into the file, and can be recorded at an opportune time.

The electronic Records file is located in the permanent data directory, under the PCR results with sub-directory. Access is restricted to certain individuals within the PCR laboratory at the discretion of the Programme Manager.

Queries and requests from individuals who submitted or intend to submit samples are frequently made. Such queries usually include:

- Advice about the nature and format of samples for submission
- Information about the outcome
- Availability of a test result
- Clinical advice may also be requested

These queries should preferably be answered by the laboratory manager. When unavailable, enquirer details including a telephone number are recorded and a note placed on the data processing table in the Reception Office.

A record of telephonic complaints and queries should be kept whenever possible.

3. Detection of Newcastle disease virus using a Triple One-step RT-PCR Assay[3]

Aim

The purpose of this assay is to routinely screen diagnostic samples. This is a rapid detection method and will assist in the identification of Newcastle disease outbreaks. This assay is applied to chicken allantoic fluid samples but can also be used on ostrich tracheal swabs. RNA is extracted from the samples according to the protocol described in section 3.8.2 of this document. The other components of the triple one-step RT-PCR are used to differentiate between virulent and avirulent strains of NDV.

(Primers ALLs/ALLe/VLTe/AVLe)

3.1. Primer sequences

ALLs	5'- TTG ATG GCA GGC CTC TTG C –3'
ALLe	5'- GGA GGA TGT TGG CAG CAT T –3'
VLTe	5'- AGC GT(C/T) TCT GTC TCC T –3'
AVLe	5'- G(A/G)C G(A/T)C CCT GT(C/T) TCC C –3'

Type	Primer set	Amplicon size
Newcastle – group	ALLs + ALLe	362 bp
Newcastle – virulent	ALLs + VLTe	254 bp
Newcastle - avirulent	ALLs + AVLe	254 bp

3.2. Responsible personnel

3.3. Safety

Newcastle disease virus is highly infectious for birds with a mildly zoonotic potential (conjunctivitis in humans). Hazardous or potentially hazardous samples must be processed in the laminar flow or Class II Biohazard cabinets. Correct protective gear e.g., lab coats and gloves must be worn at all times during the handling of Newcastle disease virus samples (see Safety Manual).

[3] Courtesy of Onderstepoort Veterinary Insititute, South Africa

3.4 Accompanying documents

Relevant SOP number:

- Primer information
- Articles
- Solution and reagent preparation
- Working instructions manual SOP
- Safety procedure document
- Safety Manual
- Sample reception SOP
- Dissemination of results

3.5. Background principle

Viral RNA is extracted from sample material, then converted into cDNA using reverse transcriptase and amplified thereafter by a Taq DNA polymerase (one-step RT-PCR).

Three tests can be done using three primer pairs, that will detect:

1. Presence of virus
2. Presence of virulent virus
3. Presence of avirulent virus

The amplified DNA is then visualised by size fractionation on an agarose gel and the results recorded by using a photo documentation system.

3.6. Pre-conditions

All laboratory personnel must be familiar with the extraction and handling of RNA, and all procedures needed to perform RT-PCR.

3.7. Procedure

3.7.1. Equipment

Micropipettes: 10 µL; 100 µL; 1000 µL

Micropipette filter tips: 10 µL

100 µL

1000 µL

Microcentrifuge tubes (2.0 mL)

Heating block 55°C (e.g., Grant QBT1)

Microcentrifuge (e.g., Jouan)

Bench top microcentrifuge (e.g., Jouan, Heraeus)

Vortex apparatus (e.g., Heidolph)

Biohazard flow cabinet (N/A 6 ft LF – Class II)

Laminar flow cabinets (e.g., Labotec)

Thermocycler (e.g., Perkin-Elmer 9700; 9600; 2400)

Microwave oven (e.g., Daewoo)

Gel apparatus (e.g., EC Electrophoretic)

Gel analyser (e.g., GeneGenius)

Ice machine (e.g., Scotsman)

3.7.2. Material and Reagents

dNTPs 10 mM (Lab. Spec.)
Oligonucleotides – ALLs/ALLe/VLTe/AVLe 20 pmol/ µL
DNA polymerase:
250 U HotStar Taq DNA polymerase
500 U FastStart Taq DNA polymerase
MMLV-RT 200 U / µL (e.g., Promega)
RNasin Ribonuclease Inhibitor 40 U/µl (e.g., Promega)
Chloroform (e.g., Merck)
Absolute ethanol (e.g., Merck)
Isopropanol (e.g., Ass. Chem. Enterprises)
70% Ethanol
DEPC-H_2O
RNase-free H_2O (e.g., USB)
1 x TAE buffer
Agarose (e.g., Separations)
DNA molecular weight marker
Loading buffer
QiaAmp Viral RNA Isolation Kit (Qiagen)
250 reactions
TRI Reagent (Sigma)

3.7.3. Control Reference Material

Test controls.

3.7.3.1. Positive controls

- Allantoic fluid from embryonated SPF chicken eggs infected with NDV (see preparation protocol)
- A high titre positive sample
- A low titre positive sample (or the above sample suitably diluted)
- NDV cDNA
- Tracheal swabs from chickens/ostriches infected with NDV (if available)

3.7.3.2. Negative controls

- Allantoic fluid from NDV-negative embryonated SPF chicken eggs.
- DEPC-H_2O
- Allantoic fluid from embryonated SPF chicken eggs infected with another virus, e.g., avian influenza (if available)

3.8. Performance of assay

3.8.1. Sample preparation

The correct protective gear e.g., lab coats and gloves must be worn at all times during sample preparation (see Safety manual).

3.8.1.1.Ostrich tracheal swabs

1. Collect a copy of the registration form on the sample reception board
2. Collect samples from -70°C freezer
3. Leave at room temperature (18-25°C) in Lab 2 to thaw
4. Pool 5 samples derived from the same farm into one microcentrifuge tube, using 50 μL from each
5. Return original samples to -70°C freezer
6. Proceed with either the TRI Reagent or QiaAmp extraction method

3.8.1.2. For other NDV samples (allantoic fluid)

1. Collect a copy of the registration form on the sample reception board
2. Collect samples from the -70°C freezer
3. Leave at room temperature (18-25°C) in Lab 2 to thaw
4. Use 250 μL from each sample separately
5. Return original samples to -70°C freezer
6. Proceed with either the TRI Reagent or QIAamp extraction method

3.8.2. RNA Isolation

3.8.2.1. TRI Reagent

1. Add 1mL of TRI Reagent to 250 µL allantoic fluid in a microcentrifuge tube
2. Leave on ice for 5 min
3. Add 200 µL of chloroform and vortex.
4. Centrifuge at 18000 g for 15 min at 4°C
5. Transfer the upper aqueous phase to a clean microcentrifuge tube and add 600 µL of isopropanol
6. Precipitate at –20°C for a minimum time of 20 min or overnight
7. Centrifuge at 18000 g for 15 min at 4°C and discard the supernatant
8. Add 900 µL of 70% ethanol and invert gently to wash the pellet
9. Centrifuge at 18000 g for 10 min at 4°C
10. Discard the supernatant and air-dry the pellet in a laminar flow cabinet for 10 min. Do not let the RNA pellet dry completely
11. Dissolve the pellet in 30 µL DEPC-H_2O / RNase-free H_2O
12. Incubate at 55°C for 10 min and store at -20°C

3.8.2.2. *QIAamp extraction kit*

1. Add 560 µL prepared AVL/Carrier RNA to 140 µL allantoic fluid in a microcentrifuge tube
2. Mix by pulse-vortexing for 15 s and incubate at room temperature (18-25°C) for 10 min
3. Add 560 µL absolute ethanol and vortex
4. Briefly centrifuge to remove drops from the inside of the lid
5. Carefully apply 630 µL of the solution to the spin column (in a 2 collection tube) and centrifuge at 6000 g / 8000 rpm for 1 min
6. Place the spin column into a clean 2 mL collection tube and discard the previous collection tube containing the filtrate
7. Repeat the previous two steps with the remaining solution
8. Add 500 µL AW1 buffer to the spin column and centrifuge at 6000 g/8000 rpm for 1 min
9. Place the spin column into a clean 2 mL collection tube and discard the previous collection tube containing the filtrate
10. Add 500 µL AW2 buffer to the spin column and centrifuge at 20 000 g /14 000 rpm (at full speed) for 3 min
11. Place the spin column into a clean microcentrifuge tube and discard the previous collection tube containing the filtrate
12. Add 60 µL elution buffer and incubate at room temperature (18-25°C) for 1 min.
13. Centrifuge at 6000 g/8000 rpm for 1 min
14. Discard the spin column and store the RNA at -20°C

3.8.2.3. Preparation of solutions

QIAamp Reagents

Carrier/RNA

1. Add 1 mL AVL buffer to Carrier RNA
2. Dissolve Carrier RNA thoroughly and transfer to the AVL buffer bottle
3. Make 1 mL aliquots and store at 4°C
4. Incubate at 80°C in a heating block for not more than 5 min before use

AW1 and AW2 buffer

Add absolute ethanol to concentrated buffer as indicated in the QIAamp manual supplied.

3.9. RT-PCR Method

3.9.1. Master mix (x1)

1. 2.5 µL 10 x HotStar or FastStart buffer
2. 1 µL dNTPs (10 mM) (see solution and reagent preparations manual)
3. 0.5 µL each of ALLe and ALLs (20 pmol/µL) primers for the virus detection reaction, and 0.5 µL each of ALLs and AVLe (20 pmol/µL) primers for avirulent virus detection, and 0.5 µL each of ALLs and VLTe (20 pmol/µL) for virulent virus detection
4. 0.05 µL RNasin Ribonuclease Inhibitor (40 U/µL); µL MMLV-RT (200 U/µl)
5. 15 µL H_2O
6. 0.25 µL or 0.5 U Taq polymerase (HotStar Taq or FastStart Taq)
7. 5 µL of isolated viral RNA

3.9.2. cDNA synthesis

Reaction is carried out in a Perkin Elmer GeneAmp PCR 9700/9600/2400 system at 42°C for 40 min, this is followed by inactivation of RTase at 95°C for 7 min.

3.9.3. PCR amplification

Denaturation	94°C	30 s
Annealing	53°C	30 s
Extension	72°C	30 s
Number of cycles		55

3.9.4. Gel separation

1. A 2 % agarose gel stained with ethidium bromide is used (Appendix B)
2. 10 μL of the product is loaded with 2 μL loading buffer
3. 2 μL of a 100 bp DNA molecular weight marker is loaded with 2 μL loading buffer in a single outside well
4. Gel electrophoresis is performed at 100 to 120 V for 30 min
5. Analysis is done by using an automated photo documentation system

3.10. Analysis and evaluation

- The positive control should yield an amplicon/fragment size of 362 bp for the group specific NDV
- For the detection of the virulent and avirulent strains an amplicon/fragment size of 254 bp is expected
- The negative control should not yield any amplicons
- The results are interpreted by accredited personnel. The case history and virus isolation results, where available , are also taken into consideration
- Results obtained are documented together with the registration forms in the specific assay file (see Section 2). Record keeping is the responsibility of the Programme Manager
- Recorded data are collated and the clients notified (farmer, state veterinarian or government office) once the laboratory Manager is satisfied that the data obtained is sufficiently reliable
- A faxed copy followed by a mail copy are sent to the clients
- The report contains the laboratory record number, the farm registration number and the outcome of the assay
- In certain cases, when requested by the submitting veterinarian, a specific report is sent containing client information and test results obtained

- An explanation of the results and/or any recommendations for future testing may be included where warranted
- This report is printed out on a letterhead, signed by the laboratory Manager and an electronic or hard copy is stored
- The copy of the report is then faxed through to the veterinarian within 12 hours after results are obtained. The letter is also sent by mail the following day. A copy of the report is kept in the appropriate file in the Receptionist's office
- A computer-based software program can be used to register arrival of samples at the laboratory and also for the sending out of results or accounts to clients. This is the responsibility of the receptionist

3.10.1. Evaluation and dissemination of results

Analytical specificity	=	100%
Analytical sensitivity	=	1.2×10^{-5} HA units (QIAamp) 6.7×10^{-3} HA units (TRI Reagent)
Diagnostic specificity	=	Only positive panels were received and analysed.
Diagnostic sensitivity	=	96%

The sensitivity was determined by a 10-fold serial dilution of virulent virus containing allantoic fluid with a titre of 1:256 HA units.

A 5 µL aliquot of 60 µl RNA extracted from 140 µl of allantoic fluid and then diluted 10^7 fold, equivalent to approximately 1.2×10^{-5} HA units, was found to be the minimum level detectable by this protocol.

3.10.2. Variations

Inhibition of the RT-PCR assay can occur on account of the following conditions:

- Inhibitors in the RT-PCR reaction (protein) - Repeat the sample extraction
- Rnase - Request new sample
- Faulty primer design - Order new primers.
- Inappropriate [Mg^{2+}] concentration - Repeat the PCR with a higher [Mg^{2+}] concentration

3.10.3. Comments

3.11. References

Wang, Z., Vreede. F.T., Mitchell, J.O. and Viljoen, G.J. 2001. Rapid detection and differentiation of Newcastle disease virus isolates by a triple one-step RT-PCR. Onderstepoort Journal of Veterinary Research, 68, 131-134.

3.12. Addendum

Articles, Primer information, RT-PCR assay performance sheet

4. The detection of *Brucella* species and strains using a PCR assay[4]

Aim

This is a rapid detection method and differentiate between Brucella abortus, Brucella suis, Brucella ovis, Brucella melitensis and Brucella abortus vaccine strains S19 and RB51. This assay is applied to whole blood, organ, foetus, semen, fluid and milk samples. DNA is extracted from the samples according to the protocol described.

(Primers ISP1/ISP2/IS711/AB/SV/OV/BM/ERI1/ERI2/RB51)

4.1. Primer sequences

ISP1	5’- GGT TGT TAA AGG AGA ACA GC –3’
ISP2	5’- GAC GAT AGC GTT TCA ACT TG –3’
IS711	5’- TGC CGA TCA CTT AAG GGC CTT CAT –3’
AB	5’- GAC GAA CGG AAT TTT TCC AAT CCC –3’
SV	5’- GCG CGG TTT TCT GAA GGT TCA GG –3’
OV	5’- CGG GTT CTG GCA CCA TCG TCG –3’
BM	5’- AAA TCG CGT CCT TGC TGG TCT GA –3’
ERI1	5’- GCG CCG CGA AGA ACT TAT CAA –3’
ERI2	5’- CGC CAT GTT AGC GGC GGT GA –3’
RB51	5’- CCC CGG AAG ATA TGC TTC GAT CC –3’

Type	Primer set	Amplicon size
Brucella group specific	ISP1 + ISP2	600 bp
B. abortus	IS711 + AB	498 bp
B. suis	IS711 + SV	285 bp
B. ovis	IS711 + OV	976 bp
B. melitensis	IS711 + BM	731 bp
B. abortus vaccine strain S19	ERI1 + ERI2	178 bp
B. abortus vaccine strain RB51	IS711 + RB51	364 bp

4.2. Responsible Personnel

All researchers in the PCR laboratory

---

[4] Courtesy of Onderstepoort Veterinary Institute, South Africa.

4.2.1. Safety

Hazardous or potentially hazardous samples must be processed in the laminar flow or Class II Biohazard cabinets. Correct protective gear e.g., lab coats and gloves must be worn at all times during the handling of *Brucella* samples (see Safety Manual).

4.3. Accompanying documents

- Relevant SOP no:
- Primer information
- Articles
- Solution and reagent preparation
- Safety procedure document
- Safety manual
- Sample reception SOP
- Dissemination of results

4.4. Background principle

DNA is extracted from sample material, and amplified by Taq DNA polymerase.
Seven tests can be done using seven sets of primers that will detect:

1. Presence of Brucella species and strains
2. Presence of Brucella abortus
3. Presence of Brucella suis
4. Presence of Brucella ovis
5. Presence of Brucella melitensis
6. Presence of Brucella abortus vaccine strain S19
7. Presence of Brucella abortus vaccine strain RB51

The amplified DNA is then visualized by size fractionation on an agarose gel and the results recorded by using a photo documentation system.

4.5. Preconditions

All laboratory personnel must be familiar with the extraction and handling of DNA, and all procedures needed to perform PCR.

4.6. Procedure

4.6.1. Equipment

Micropipettes: 10 µL; 100 µL; 1000 µL (e.g., Eppendorf)

Micropipette filter tips	10 µL
	100 µL
	1000 µL

Microcentrifuge tubes (1.5 mL)

Heating block 50°C and 100°C (e.g., Grant QBT1)

Microcentrifuge (e.g., Jouan)

Bench top microcentrifuge (e.g., Jouan, Heraeus)

Vortex apparatus (e.g., Heidolph)

Biohazard flow cabinet (N/A 6 ft LF)

Laminar flow cabinets (e.g., Labotec)

Thermocycler (e.g., Perkin-Elmer 9700; 9600; 2400)

Microwave oven (e.g., Daewoo)

Gel apparatus (e.g., EC Electrophoretic)

Gel analyser (e.g., GeneGenius)

Ice machine (e.g., Scotsman)

4.6.2. Material and Reagents

dNTPs 10 mM (e.g., Lab. Spec.)

Oligonucleotides

ISP1/ISP2/IS711/AB/SV/OV/BM/ERI1/ERI2/RB51

20 pmol/μL (e.g., Whitehead)

DNA polymerase:

500 U Promega Taq DNA polymerase

500 U FastStart Taq DNA polymerase

Absolute ethanol (e.g., Merck)

70% Ethanol

Biophenol (e.g., Biosolve)

Buffer # 1

Buffer # 2

1 x SSC

30% Hydrogen peroxide

Proteinase K (20 mg/mL)

7.5 M Ammonium acetate

1 x TAE buffer

Agarose (e.g., Separations)

DNA molecular weight marker

Loading buffer

4.6.3. Control Reference Material

Test controls:

4.6.3.1. Positive controls

DNA from *B. abortus*, *B. suis*, *B. ovis*, *B. melitensis* and *B. abortus* vaccine strains S19 and RB51 positive samples

- A high titre positive sample
- A low titre positive sample (or the above sample suitably diluted)

4.6.3.2. Negative controls

DNA from *B. abortus*, *B. suis*, *B. ovis*, *B. melitensis* and *B. abortus* vaccine strains S19 and RB51 negative samples.

- DEPC-H_2O

4.7. Performance of assay

4.7.1. Sample preparation

The correct protective gear e.g., lab coats and gloves must be worn at all times during sample preparation.

4.7.2. Whole blood, organ, foetus, semen, fluid and milk samples

- Collect a copy of the registration form on the sample reception board
- Collect samples from 4°C fridge
- Proceed with the Solution D / Chomczynski extraction method

4.8. DNA Isolation (Solution D/Chomczynski extraction)[5]

4.8.1. Whole blood, semen, organs, foetus and milk samples.

Add 500 µL of Buffer # 1 to 500 µL of the sample.

[5] See Chapter 7, Section 6.1.

4.8.2. Organ and foetus samples

1. Homogenise approximately 0.25 g of the sample in 1.5 mL of Buffer # 1
2. Vortex
3. Incubate on ice for 30 min
4. Centrifuge at 12000 rpm for 15 min
5. Decant supernatant
6. Add 500 µL of 1 x SSC and vortex
7. Centrifuge at 12000 rpm for 15 min
8. Decant supernatant

4.8.3. Additional step for extraction of whole blood samples

1. Add 100 µL of 30% hydrogen peroxide to the pellet
2. Incubate for 2-5 min at room temperature (18-25°C) without disturbing the pellet
3. Remove the supernatant with a pipette
4. Wash the pellet with 1 mL H_2O and remove the H_2O without centrifugation
5. Add 500 µL of Buffer # 2 and vortex
6. Add 20 µL of Proteinase K
7. Incubate for 2 hours to overnight at 50°C
8. Boil the sample at 100°C for 3 min
9. Add 500 µL of Biophenol and vortex
10. Centrifuge at 12000 rpm for 10 min
11. Transfer the upper clear phase to a clean tube
12. Add 3 µL of 7.5 M ammonium acetate
13. Add 2 volumes of absolute ethanol (approximately 900 µL)
14. Centrifuge at 12000 rpm for 10 min
15. Decant supernatant
16. Add 200 µL of 70% ethanol to wash the pellet
17. Centrifuge at 12000 rpm for 10 min
18. Discard supernatant
19. Air dry pellet for approximately 10-15 min
20. Dissolve the pellet in 50 µL of HPLC water
21. Store at -20°C

Dilute DNA according to the sample type before use in PCR reaction

Whole blood samples	Use undiluted
Semen, fluid and milk samples	1:5
Organ samples	1:10

4.8.4. DNA Extraction from Milk/Semen with TRI REAGENT® LS (Method 1)

4.8.4.1. Product description

TRI Reagent®LS is a ready to use reagent for the simultaneous isolation of RNA, DNA and proteins from liquid samples of human, animal, plant, yeast, bacterial and viral origin. TRI Reagent LS combines phenol and guanidine thiocyanate in a mono-phase solution. A biological sample is homogenized or lysed in TRI Reagent LS and the lysate is separated into aqueous and organic phases by bromochloropropane (cat. no. BP 151) or chloroform addition and centrifugation. RNA remains exclusively in the aqueous phase, DNA in the interphase and proteins in the organic phase. RNA is precipitated from the aqueous phase by the addition of isopropanol and the RNA pellet is washed with ethanol and solubilized. DNA and proteins are sequentially precipitated from the interphase and organic phase with ethanol and isopropanol, washed with ethanol and solubilized.

4.8.4.2. Isolation of DNA from Milk/ Semen Protocol

The DNA is isolated from the interphase and phenol phase separated from the initial homogenate as described above. Following precipitation and a series of washes, the DNA is solubilized in 8 mM NaOH, neutralized and used for analysis. The DNA isolated by TRI Reagent LS can be used for PCR, restriction digestion and Southern blotting.

The protocol includes the following steps:

1. Homogenisation	0.75 mL TRI Reagent LS + 0.25 mL sample.
2. Phase Separation	homogenate + 0.1 mL BCP or 0.2mL chloroform.
3. DNA Precipitation	phenol phase and interphase + 0.3 mL ethanol (0.75 TRI Reagent LS).
4. DNA Wash:	1mL 0.1 M trisodium citrate in 10% ethanol x 2; 2 mL 75% ethanol.
5. DNA Solubilisation	8 mM NaOH.

The procedure is carried out at room temperature, unless stated otherwise.

1. Homogenisation

Biological fluids. Mix 0.75 mL of TRI Reagent LS with 0.25 mL of sample and lyse cells (or cellular debris) suspended in the sample by passing the suspension several times through a pipette.

If the sample volume is < 0.25 mL, adjust the volume to 0.25 mL with water. The volume ratio of TRI Reagent LS to sample should be 3 to 1.

2. Phase Separation

Store the lysates/homogenates for 5 min at room temperature to permit the complete dissociation of nucleoprotein complexes. Next, supplement the lysate with 0.2 mL chloroform per 0.75 mL of TRI Reagent LS, cover the samples tightly and shake vigorously for 15 s. Store the resulting mixture at room temperature for 2-15 min. Centrifuge the resulting mixture at 12,000 g for 15 min at 4°C. Following centrifugation, the mixture separates into a lower red phenol-chloroform phase, interphase and the colorless upper aqueous phase. RNA remains exclusively in the aqueous phase whereas DNA and proteins are in the interphase and organic phase. The volume of the aqueous phase is about 70% of the volume of TRI Reagent LS used for homogenization.

Chloroform used for phase separation should not contain isoamyl alcohol or any other additive. It is important to perform centrifugation to separate aqueous and organic phases in the cold (4-10°C). If performed at elevated temperature, a residual amount of DNA may sequester in the aqueous phase.

3. DNA Precipitation

Remove the remaining aqueous phase overlying the interphase and precipitate DNA from the interphase and organic phase with ethanol. Add 0.3 mL of 100% ethanol per 0.75 mL of TRI Reagent LS used for the initial homogenization and mix samples by inversion. Next, store the samples at room temperature for 2-3 min and sediment DNA by centrifugation at 2,000 g for 5 min at 4°C. Careful removal of the aqueous phase is critical for the quality of the isolated DNA.

If the DNA yield is expected to be low (titre of organisms very low) implement an alternative DNA isolation procedure as follows.
Remove the remaining aqueous phase overlying the interphase. Add 2-8 uL of Polyacryl Carrier (see footnote) to the interphase- phenol phase. Add 0.3 mL of 100% ethanol per 0.75 mL of TRI Reagent LS used for the initial homogenization and mix samples by inversion. Next, store the samples at room temperature for 2-3 min and sediment DNA by centrifugation at 2,000 g for 5 min at 4°C. Now follow the alternative DNA wash **(as described in 4 below)**

4. DNA Wash

Remove the phenol-ethanol supernatant. Wash the DNA pellet twice in a solution containing 0.1 M trisodium citrate in 10% ethanol. (No pH adjustment required.) Use 1 mL of the solution per 0.75 mL of TRI Reagent LS used for the initial homogenization. At each wash, store the DNA pellet in the washing solution for 30 m at room temperature with periodic mixing and centrifuge at 2,000 g for 5 m at 4 - 25 C. Following these two washes, suspend the DNA pellet in 75% ethanol (1.5 - 2 mL of 75% ethanol per 0.75 mL TRI Reagent LS), store for 10-20 min at room temperature with periodic mixing and centrifuge at 2,000 g for 5 m at 4-25°C . This ethanol wash removes pinkish color from the DNA pellet. An additional wash in 0.1M trisodium citrate-10% ethanol is required for large pellets containing >200 μg DNA or large amounts of a non-DNA material.

Alternative DNA wash

Replace the sodium citrate washes described above by performing 2 x 10 min washes of the DNA/carrier pellet using 75% ethanol with intermittent mixing. Proceed with the rest of the protocol as described below.

5. DNA Solubilisation

Remove the ethanol wash and briefly air-dry the DNA pellet by keeping tubes open for 3-5 min at room temperature. Dissolve the DNA pellet in 8 mM NaOH by slowly passing through a pipette. Add an adequate amount of 8 mM NaOH to approach a DNA concentration of 0.2 - 0.3 μg/μl. Typically add 0.3-0.6 mL 8 mM NaOH to DNA isolated from 10^7 cells. The use of a mild alkaline solution assures full solubilization of the DNA pellet. At this stage, the DNA preparations may contain insoluble material (fragments of membranes, etc.). Remove insoluble material by centrifugation at 12,000 g for 10 min and transfer the resulting supernatant containing DNA to a new tube. A high viscosity of the supernatant indicates the presence of high molecular weight DNA.

6. Quantification OF DNA (optional)

For optimal spectrophotometric measurements, DNA aliquots should be diluted with water or buffer with a pH >7.5 such as Phosphate Buffer. Distilled water with a pH < 7.0 falsely decreases the 260/280 ratio and impedes the detection of protein in RNA samples. Calculate the DNA content assuming that one A260 unit equals 50 μg of double-stranded DNA/mL.

Amplification of DNA by PCR

Following solubilization in 8 mM NaOH, adjust the pH of the DNA sample to 8.4 using HEPES ***(see* Table below).** Add an aliquot of the sample (typically 0.1-1 µg DNA) to a PCR reaction mix and perform PCR according to your standard protocol.

Adjustment of pH in DNA samples solubilised in 8 mM NaOH

For 1 mL of 8mM NaOH, use the following amounts of 0.1M or 1.0 M HEPES (free acid)

Final pH	**0.1M HEPES (µL)**	**Final pH**	**1M HEPES (µL)**
8.4	86	7.2	23
8.2	93	7.0	32
8.0	101		
7.8	117		
7.5	159		

Notes

1. If necessary, the phenol phase and interphase can be stored at 4 °C overnight.
   - Samples suspended in 75% ethanol can be stored at 4°C for a long period of time (months).
   - Samples solubilized in 8 mM NaOH can be stored overnight at 4°C.
   - For prolonged storage, adjust samples to pH 7-8 and supplement with 1 mM EDTA.

2. Molecular weight of the isolated DNA depends on the shearing forces applied during homogenization. When possible, use a loosely fitting homogenizer.

3. The isolation protocol can be modified if the DNA is isolated only for quantitative purposes:

   - A more vigorous homogenization of samples can be performed, including the use of Polytron.
   - Phenol phase and interphase can be stored at 4 °C for a few days or at -70 °C for a few months.
   - Solubilization of DNA can be facilitated by replacing 8 mM NaOH with a 40 mM solution and by vortexing of the DNA pellet instead of pipetting.

Do not shorten the recommended time of storing samples with the washing solutions. These are the minimal periods of time necessary for efficient removal of phenol from the DNA pellet.

4. To assure full recovery of DNA from small samples (<10 µg DNA), we recommend the use of **Polyacryl Carrier (cat. no. PC 152).**

- Perform homogenization, phase separation and removal of the aqueous phase as described in the RNA Isolation section of the protocol.
- Remove any remaining aqueous phase overlying the interphase and add 2-8 µl of Polyacryl Carrier to the interphase-phenol phase.
- Replace the sodium citrate washes described in Step 2 by performing two 10 min washes of the DNA/carrier pellet using 75% ethanol with intermittent mixing.
- Proceed with DNA solubilization as described in the protocol.
- This alternative procedure replaces steps 1-2 of the DNA Isolation procedure.
- Prepare a back extraction buffer containing 4 M guanidine thiocyanate, 50 mM sodium citrate and 1 M Tris (free base).
- Following phase separation RNA Isolation Procedure Step 2, remove any remaining aqueous phase overlying the interphase and add back extraction buffer to the interphase-organic phase mixture.
- Use 0.5 mL of back extraction buffer per 0.75 mL of TRI REAGENT LS used for the initial homogenization. Vigorously mix the sample by inversion for 15 s and store for 10 min at room temperature.
- Perform phase separation by centrifugation at 12,000 g for 15 min at 4°C.
- Transfer the upper aqueous phase containing DNA to a clean tube and save the interphase and organic phase at 4°C for subsequent protein isolation.
- Precipitate DNA from the aqueous phase by adding 0.4 mL of isopropanol per 0.75 mL of TRI REAGENT LS used for the initial homogenization.
- Mix the sample by inversion and store for 5 min at room temperature.
- If the expected DNA yield is less than 20 µg, add 2-8 µl of Polyacryl Carrier to the aqueous phase prior to isopropanol addition and mix.
- Sediment DNA by centrifugation at 12,000 g for 5 min at 4-25°C and remove the supernatant. Wash the DNA pellet with 1.0 mL of 75% ethanol and proceed with DNA solubilization as described in Step 3.

Also see the Troubleshooting Guide.

4.8.4.3. Troubleshooting

DNA Isolation

- Low yield. a) Incomplete homogenization or lysis of samples, b) incomplete solubilization of the final DNA pellet.
- 260/280 ratio < 1.70. Phenol was not sufficiently removed from the DNA preparation. Add one more wash of the DNA pellet with 10% ethanol-0.1 M sodium citrate solution.
- DNA degradation. a) Tissues were not immediately processed or frozen after removing from animal or plant, b) samples were homogenized with a Polytron or other high speed homogenizer.
- RNA contamination. a) Too large volume of aqueous phase remained with the interphase and organic phase, b) DNA pellet was not sufficiently washed with 10% ethanol-0.1 M sodium citrate solution

4.8.4.4. Polyacryl carrier

Store at 4 °C.

Product Description

PolyAcryl Carrier is a molecular biology grade solution of acryl polymer designed for use in the isolation of small amounts of RNA or DNA. PolyAcryl Carrier does not affect the activity of restrictases, reverse transcriptase, Taq polymerase, DNA polymerase, ligase or other enzymes used for nucleic acid analysis. The carrier can be used for at least one year when stored at room temperature or 4 °C. The shelf-life can be extended beyond one year by storage at -20 °C.

Application

For the isolation of RNA or DNA using TRI Reagent® (cat. no. TR 118) or DNAzol® (cat. no. DN 127), add 2-8 µl of PolyAcryl carrier per 1.0 mL of TRI Reagent or DNAzol. Perform homogenization or lysis of the sample and isolate the RNA or DNA as described in the respective protocols. For the precipitation of nucleic acids from other solutions, add 2-8 µl of PolyAcryl carrier per 1.0 mL of RNA or DNA solution. Briefly mix the solution and carry out precipitation by adding NaCl to a final concentration of 0.2 M, followed by 2.5 volumes of ethanol. Mix the solution and store at room temperature for 5-10 min. Sediment DNA precipitate at 3,000 g or RNA precipitate at 10,000 g for 10 min at 4°C. Dissolve the precipitate by repetitive pipetting in DEPC-treated water or other solubilising medium. PolyAcryl Carrier contributes to the optical density of RNA and DNA preparations. To normalize for this effect, process a blank sample containing only the reagent and PolyAcryl Carrier. Solubilise the final pellet of the carrier in the same volume of solubilisation solution used for the test sample pellets. Measure the optical density at 260 and 280 nm for all samples and the blank, and subtract the values of the blank from values obtained for each DNA sample.

4.8.5. DNA Extraction from Milk –PCR application Method 2

Buffers and reagents

1. Buffer #1 (20mM NaCl, 20mM EDTA, 20 mM Tris-HCl pH= 7,5, 0,5%Triton)
2. Buffer #2 (10mM NaCl, 50mM EDTA, 50mM Tris-HCl pH= 7,5, 1% SDS)
3. 20X SSC (NaCl 175.3g, Sodium Citrate 88,2 g, Dissolve in 800 mL of water, Adjust pH to 7.2 by adding a few drops of concentrated HCl, then adjust volume to 1000 mL). Dilute to use in protocol as 1X SSC.
4. Proteinase K (20 mg/ml)
5. Biophenol (phenol-chloroform- isoamylalcohol)
6. Ammonium acetate 7.5M
7. Absolute ethanol/ 70% Ethanol

Procedure

1. 500 uL of sample (milk, culture, organs tissues, blood etc)
2. Add 500 uL of buffer #1
3. Leave on ice for 30 min
4. Centrifuge at 12000 rpm for 15 min at room temperature or 4 °C
5. Discard the supernatant
6. Add 500 μL of 1x of SSC
7. Vortex
8. Centrifuge at 12000 rpm for 15 min
9. Discard the supernatant
10. Add 500 μL of Buffer #2
11. Mix and vortex
12. Add 20 μL of proteinase K (20mg/ml)
13. Incubate for 2 hours at 50C (or overnight)
14. Add 500 μL of biophenol
15. Vortex
16. Centrifuge at 12000 rpm for 10 min
17. Keep the supernatant in one clean tube
18. Add 1/3 of volume (supernatant) 7.5M ammonium acetate
19. Add 2 volumes of absolute ethanol and mix
20. Centrifuge at 12000 rpm for 10 min
21. Discard the supernatant
22. Add 200 μL of 70% ethanol to wash DNA pellet
23. Centrifuge at 12000 rpm for 10 min
24. Dry pellet at room temperature
25. Dissolved the pellet in 50 μL of HLPC water
26. Store at - 20 °C

4.8.6. DNA Extraction from Milk –PCR application Method 3

Improved Method for Purification of Brucella DNA from Bovine Milk: Romero and Lopez-Goñi (1999) (Applied and Environmental Microbiology, 65, 3735-3737).

Limit of detection as low as 5-50 CFU/ml

Reagents

1. NET buffer (50 mM NaCl, 125 mM EDTA, 50 mM Tris-HCl pH 7.6)
2. 24% SDS
3. RNase
4. Proteinase K

Procedure

1. Milk can be frozen
2. Thaw at room temperature
3. Mix 500 µL milk with 100 µL NET buffer
4. Add 100 µL 24% SDS (final concentration, 3.4%)
5. Incubate at 80 °C for 10 minutes
6. Cool sample
7. Add 75 µg RNase/ml and incubate at 50C for 2 h (can give better results)
8. Add Proteinase K (final concentration of 325 µg/ml or more); incubate at 50 °C or at least 1.5 h (37 °C also works)
9. Perform standard phenol-chloroform-isoamyl alcohol extraction
10. Precipitate with isopropanol
11. Wash with 70% ethanol
12. Air-dry under vacuum
13. Dissolve pellet in 25 µL sterile dd water
14. Store at –20 °C until further use
15. 1 µL can be added to PCR cocktail

4.8.7. DNA Extraction from Milk for PCR application Method 4

Use brucella antibody coated magnetic beads to trap brucella out of milk, and then normal extraction. Nucleic acid spin columns can also be used in place of precipitation.

4.9. PCR Method

4.9.1. Master mix (x 1)

1. 2.5 µL 10 x PCR buffer
2. 0.5 µL dNTPs (10 mM)
3. 0.5 µL each of: ISP1 and ISP2 (20 pmol/µL) primers for *Brucella* group specific, IS711 and AB (20 pmol/µL) primers for *B. abortus*, IS711 and SV (20 pmol/µL) primers for *B. suis*, IS711 and OV (20 pmol/µL) primers for *B. ovis*, IS711 and BM (20 pmol/µL) primers for *B. melitensis*, ERI1 and ERI2 (20pmol/µL) primers for *B. abortus* vaccine strain S19, IS711 and RB51 (20 pmol/µL) primers for *B. abortus* vaccine strain RB51
4. 17.75 µL H_2O
5. 0.25 µL or 2U Taq polymerase (e.g. Promega Taq)
6. 3 µL of isolated DNA

4.9.2. PCR amplification for Brucella group specific

Denaturation	95°C	35 s
Annealing	56°C	45 s
Extension	72°C	45 s
Number of cycles	30	

4.9.3. PCR amplification for Brucella spp. and strains

Initial denaturation	94°C	2 min
Denaturation	95°C	20 s
Annealing	55.5°C	20 s
Extension	72°C	30 s
Number of cycles	35	

4.9.4. Gel separation

1. A 1.5 % agarose gel stained with ethidium bromide is used **(Appendix B)**
2. 10 µL of the product is loaded with 2 µL loading buffer
3. 2 µL of a 100 bp DNA molecular weight marker is loaded with 2 µL loading buffer in a single outside well
4. Gel electrophoresis is performed at 100 to 120 V for 30 min.

Analysis is done by using an automated photo documentation system.

4.10. Analysis and evaluation

The positive control should yield fragment sizes of :

600 bp for *Brucella* group specific

498 bp for *B. abortus*

285 bp for *B. suis*

976 bp for *B. ovis*

731 bp for *B. melitensis*

178 bp for *B. abortus* vaccine strain S19

364 bp for *B. abortus* vaccine strain RB51

The negative control should not show any amplicons.

The results are interpreted by accredited personnel.

Results are documented in the specific assay file, record keeping is the responsibility of the laboratory manager.

A report is sent to the submitting veterinarian containing client information and test results obtained. An explanation of the results and/or any recommendations for future testing may be included where warranted. This report is printed out on a letterhead, signed by the Division head or his deputy, and an electronic copy is stored on a permanent drive.

The copy of the report is then faxed through to the veterinarian within 12 h after results are obtained. The letter is also sent by mail the following day.

A copy of the report is kept in the appropriate file in the receptionist's office.

A computer-based software program can be used to register arrival of samples at the section and also for the sending out of accounts to clients. This is the responsibility of the receptionist.

4.10.1. Variations

Inhibition of the PCR assay can occur on account of the following conditions:

Observed	Action
Inhibitors in the PCR reaction (protein).	Repeat the sample extraction.
DNase.	Request new sample.
Faulty primer design.	Order new primers.
Inappropriate [Mg^{2+}] concentration.	Repeat the PCR with a higher [Mg^{2+}] concentration.

4.11. References

1. Ausubel F., Brent R., Kingston R., Moore D., Seidman J., Smith J. and Struhl K. 1990. Appendix 1, Current Protocols in Molecular Biology, 2, A.1.5, John Wiley and Sons, Inc., New York, NY

2. Banerjee, S, Smallwood A, Chambers AE and Nicolaides K. (2003) Quantitative Recovery of Immunoreactive Proteins from Clinical Samples Following RNA and DNA Isolation. BioTechniques, 35, 450-456

3. Bricker, B.J. and Halling, S.M. 1994. Differentiation of *Brucella abortus* bv. 1, 2 and 4, *Brucella melitensis*, *Brucella ovis* and *Brucella suis* bv. 1 by PCR. Journal of Clinical Microbiology, 32, 2660-2666

4. Bricker, B.J. and Halling, S.M. 1995. Enhancement of the *Brucella* AMOS PCR assay for differentiation of *Brucella abortus* vaccine strains S19 and RB51. Journal of Clinical Microbiology, 33, 1640-1642

5. Chomczynski P. and Sacchi N. 1987. Single step method of RNA isolation by acid guanidinium thiocyanate-phenol-chloroform extraction. Anal Biochem, 162, 156-159

6. Chomczynski P. 1993. A reagent for the single-step simultaneous isolation of RNA, DNA and proteins from cell and tissue samples. BioTechniques, 15, 532-537

7. Chomczynski P. and Mackey K. 1995. Modification of the TRI Reagent™ procedure for isolation of RNA from polysaccharide - and proteoglycan - rich sources. Biotechniques, 19, 942-945

8. Chomczynski P. and Mackey K. 1995. Substitution of chloroform with bromochloropropane in the single-step method of RNA isolation. Anal Biochem, 225, 163-164

9. Ouahrani-Bettache, S., Soubrier, M.-P. and Liautard, J.-P. 1996. IS6501-anchored PCR for the detection and identification of *Brucella* species and strains. Journal of Bacteriology, 81, 154-160

10. Mackey K. and Chomczynski P. 1996. Long-term stability of RNA isolation reagents. J NIH Res., 8, 72

11. Wilfinger W., Mackey K. and Chomczynski P. 1997. Effect of pH and ionic strength on the spectrophotometric assessment of nucleic acid purity. BioTechniques, 22, 474-481

12. Wu, LC (1997) Isolation and Long-Term Storage of Proteins from Tissues and Cells Using TRIzol Reagent. FOCUS, 17, 98-100

5. The detection of Malignant catarrhal fever virus using a PCR assay[6]

Aim

This is a rapid detection method and differentiate between sheep-associated and wildebeest-associated malignant catarrhal fever. This assay is applied to whole blood, blood clot and organ samples.

(Primers SNF1/SNR1/SNF2/SNR2/WBN1/WR)

5.1. Primer sequences

SNF1	5'- GTA TCC GAA AGC AGC CCC AGT ATC –3'
SNR1	5'- ACA GCT GGG GCA GGA TTA CAG AC –3'
SNF2	5'- AGC ACA GTT TAT TTC AGA C –3'
SNR2	5'- GAT AAG CAC CAG TTA TGC –3'
WBN1	5'- CGT ACC CAC TGG GTA AG –3'
WR	5'- GGC CTA TCC TAT AAG AC –3'

Type	**Primer set**	**Amplicon size**
Sheep-associated MCF	SNF1 + SNR1 + SNF2 + SNR2	274 bp
Wildebeest-associated MCF	WBN1 + WR	241 bp

5.2. Accredited Personnel

All PCR laboratory personnel.

[6] Courtesy of Onderstepoort Veterinary Research Institute, South Africa.

5.3. Safety

Hazardous or potentially hazardous samples must be processed in the laminar flow or Class II Biohazard cabinets. Correct protective gear e.g., lab coats and gloves must be worn at all times during the handling of malignant catarrhal fever samples.

5.4. Accompanying documents

Primer information

Articles

Solution and reagent preparation

Safety procedure document

Sample reception SOP

Dissemination of results

5.5. Background principle

DNA is extracted from sample material, and amplified by Taq DNA polymerase.

Two tests can be done using two sets of primers, which will detect:

1. Presence of the sheep-associated agent of malignant catarrhal fever

2. Presence of the wildebeest-associated agent of malignant catarrhal fever

The amplified DNA is then visualised by size fractionation on an agarose gel and the results recorded by using a photo documentation system.

5.6. Preconditions

All laboratory personnel must be familiar with the extraction and handling of DNA, and all procedures needed to perform PCR.

5.7. Procedure

5.7.1. Equipment

Micropipettes: 10 µL; 100 µL; 1000 µL (e.g., Eppendorf)

Micropipette filter tips: 10 µL

100 µL

1000 µL

Microcentrifuge tubes (1.5 mL)

Microcentrifuge (e.g., Jouan)

Bench top microcentrifuge (e.g., Jouan, Heraeus)

Vortex apparatus (e.g., Heidolph)

Biohazard flow cabinet (N/A 6 ft LF)

Laminar flow cabinets (e.g., Labotec)

Thermocycler (e.g., Perkin-Elmer 9700; 9600; 2400)

Microwave oven (e.g., Daewoo)

Gel apparatus (e.g., EC Electrophoretic)

Gel analyser (e.g., GeneGenius)

Ice machine (e.g., Scotsman)

5.7.2. Material and Reagents

dNTPs 10 mM (e.g., Lab. Spec.)

Oligonucleotides – SNF1/SNR1/SNF2/SNR2/WBN1/WR

20 pmol/μL (e.g., Whitehead)

DNA polymerase:

500 U Promega Taq DNA polymerase

500 U FastStart Taq DNA polymerase

Chloroform (e.g., Merck)

Absolute ethanol (e.g., Merck)

-70% Ethanol

DEPC-H_2O

1 x TAE buffer

Agarose (e.g., Separations)

DNA molecular weight marker

Loading buffer

Nucleon BACC2 Extraction and Purification kit (e.g., Amersham Pharmacia)

5.7.3. Control Reference Material

Test controls

5.7.3.1. Positive controls

- DNA from MCF-SA and WB positive field samples
- A high titre positive sample
- A low titre positive sample (or the above sample suitably diluted)

5.7.3.2. Negative controls

DNA from MCF-SA and WB negative field samples
DEPC-H_2O

5.8. Performance of assay

5.8.1. Sample preparation

The correct protective gear e.g., lab coats and gloves must be worn at all times during sample preparation.

5.8.2. Blood and organ samples

Collect a copy of the registration form on the sample reception board

Collect samples from 4°C refrigerator

Proceed with the Amersham Nucleon BACC2 extraction method

5.8.3. DNA Isolation

5.8.3.1. Whole blood sample

1. Add 800 µL of Solution A to 300 µL of the sample
2. Incubate at room temperature (18-25°C) for 4 min, rotary mixing the tube constantly
3. Centrifuge at 1300 g for 5 min

5.8.3.2. Blood clot / organ sample

1. Homogenise approximately 0.25g of the sample in 2.5 mL of Reagent A of Nucleon BACC2 extraction kit
2. Incubate at room temperature (18-25°C) for 4 min, rotary mixing the tube constantly
3. Centrifuge at 1300g for 10 min
4. Decant supernatant
5. Add 350 µL of Solution B and vortex
6. Add 100 µL of sodium perchlorate solution
7. Invert the tube 25-30 times
8. Add 600 µL of chloroform and vortex
9. Add 100 µL of resin. Do not shake the tube
10. Centrifuge at 300 g for 3 min
11. Collect the upper clear phase (approximately 450 µL) and transfer to a clean tube
12. Add 900 µL of cold absolute ethanol. Mix by inversion until the precipitate appears
13. Centrifuge at 13000 rpm for 5 min
14. Decant supernatant
15. Add 500 µL of 70% ethanol to the pellet and vortex
16. Centrifuge at 13000 rpm for 5 min
17. Decant supernatant
18. Dry the pellet for 20 min, ensuring that all the ethanol has been removed
19. Dissolve in 200 µL of HPLC water and store at -20°C

5.9. PCR Method

5.9.1. Master mix for MCF-SA (x 1)

a. 2.5 μL 10 x Dynazyme buffer (see solution and reagent preparations manual)
b. 0.5 μL dNTPs (10 mM)
c. 1 μL each of SNF1 (1:10), SNR1 (1:10), SNF2 and SNR2 (20 pmol/μl) primers
d. 12.75μL H_2O
e. 0.25 μL or 2 U Taq polymerase (e.g., Promega Taq)
f. 5 μL of isolated DNA

5.9.2. Master mix for MCF-WB (x1)

a. 1.25 μL 10 x FastStart buffer
b. 0.25 μL FastStart $MgCl_2$
c. 0.25 μL dNTPs (10 mM)
d. 0.5 μL each of WBN1 and WR (20 pmol/ μL) primers
e. 9.15 μL H_2O
f. 0.1 μL FastStart Taq polymerase
g. 1 μL of isolated DNA

5.9.3. PCR amplification for MCF-SA

Initial denaturation	93°C	2 min
Denaturation	94°C	30 s
Annealing	68°C	30 s
Extension	72°C	30 s
Number of cycles	20	

Denaturation	94°C	30 s
Annealing	54°C	30 s
Extension	72°C	30 s
Number of cycles	30	

5.9.4. PCR amplification for MCF-WB

Initial denaturation	95°C	7 min
Denaturation	94°C	30 s
Annealing	54°C	30 s
Extension	72°C	30 s
Number of cycles	50	

5.10. Gel separation

1. A 1.5 % agarose gel stained with ethidium bromide is used
2. 10 µL of the product is loaded with 2 µL loading buffer
3. 2 µL of a 100 bp DNA molecular weight marker is loaded with 2 µL loading buffer in a single outside well
4. Gel electrophoresis is performed at 100 to 120 V for 30 min
5. Analysis is done by using an automated photo documentation system

5.11. Analysis and evaluation

The positive control should yield a fragment size of:

274 bp for the sheep-associated MCF
241 bp for the wildebeest-associated MCF

The negative control should not show any amplicons.
The results are interpreted by accredited personnel by looking at the case history and correlating the results with previous results.

Results are documented in the specific assay file. Record keeping is the responsibility of the Technical Manager.

A report is sent to the submitting veterinarian containing client information and test results obtained. An explanation of the results and/or any recommendations for future testing may be included where warranted.

This report is printed out on a letterhead, signed by the Division head or his deputy, and an electronic copy is stored on the local network (H-drive). The copy of the report is then faxed through to the veterinarian within 12 h after results are obtained. The letter is also sent by mail the following day. A copy of the report is kept in the appropriate file in the Receptionist's office.

A computer-based software programme (Progress) is used to register arrival of samples at the section and also for the sending out of accounts to clients. This is the responsibility of the Receptionist.

Evaluation and dissemination of results **(This chapter, Section 2)**

5.12. Variations

Inhibition of the PCR assay can occur on account of the following conditions:

Observation	**Action**
Inhibitors in the PCR reaction (protein)	Repeat the sample extraction
Dnase	Request new sample
Faulty primer design	Order new primers
Inappropriate [Mg^{2+}] concentration	Repeat the PCR with a higher [Mg^{2+}] concentration.

5.13. References

1. Baxter, S.I.F., Pow, I., Bridgen, A. and Reid, H.W. 1993. PCR detection of the sheep-associated agent of malignant catarrhal fever. Archives of Virology, 132, 145-159

2. Ensser, A., Pflanz, R. and Fleckenstein, B. 1997. Primary structure of the Alcelaphine Herpesvirus 1 genome. Journal of Virology, 71, 6517-6525

6. Detection of Foot and mouth disease virus, Rinderpest virus, Peste-des-petits-ruminants virus and Bovine viral diarrhoeal virus using a RT-PCR assay[7]

These are major diseases of livestock and their diagnosis is vital in control programmes. PCR is used in tandem with a series of serological techniques to determine the presence of nucleic acid in samples. These are RNA viruses so that adaptations to the described PCRs so far are necessary.

6.1.Preparative solutions and reagents

6.1.1. Solution D

The procedure is that recommended by Chomczynski and Sacchi 1987 which minimises the hazard of handling the poisonous guanidium thiocyanate.

It should be carried out in a chemical safety hood.

The following are the amounts for a 250g bottle but the volumes can be adjusted for the other quantities available from the supplier (Fluka).

1. Dissolve the guanidium thiocyanate in the manufacturer's bottle in 293 mL sterile distilled water, 17.6 mL. 0.75M sodium citrate, pH 7.0, and 26.4 mL 10% sarcosyl

2. Heat to 65°C in a water bath to dissolve

3. This can be kept for several months in the dark at room temperature in a chemical safety cabinet

4. The final solution D is made by the addition of 0.36 mL of 2-mercaptoethanol to 50 mL of the stock solution and this should not be kept for more than one month so it is best to prepare just enough for a month's needs

6.1.2. Phenol

To avoid the hazard of preparing pure water-saturated and buffer-saturated phenol for RNA extraction it is best to buy them ready prepared from a chemical supplier (e.g., Gibco-BRL).

6.1.3. Chloroform/isoamyl alcohol

Mix the two in a ratio of 49 parts chloroform to 1 part isoamyl alcohol.

---

[7] Courtesy of Institute for Animal Heath, UK., and the National Veterinary Institute, Sweden.

6.1.4. 2M sodium acetate

Weigh out the correct amount of acetate and add about quarter the amount of distilled water needed to make the final concentration. Add acetic acid until the pH is 4.2 (Note: it takes quite a lot of acetic acid to alter the pH) and make up to the final volume with distilled water. Autoclave in small aliquots (10-20 mL).

6.2. RNA extraction from diagnostic samples

To detect virus-specific RNA in diagnostic specimens the samples must be in reasonably good condition since the RNA genome is easily destroyed by heat or enzymes released during putrefaction.

Infected tissues will contain large amounts of virus nucleocapsids, i.e., virus genome RNA wrapped up in a protecting protein shell. This helps to protect the RNA from total degradation by the powerful RNases released when cells are damaged and, while viable virus may not be obtained from all infected tissues, molecular techniques can be used to detect the presence of virus-specific RNA even if the virus is non-viable. The method chosen to purify the RNA from the tissue samples should rapidly inactivate the RNase released when the tissues are further damaged by the homogenisation procedure. The Acid-guanidinium-phenol-chloroform (AGPC) is a rapid and simple method that is suitable for extracting RNA from any type of tissue sample. Note: Field samples should be handled with gloves and opened and processed in a biological safety hood until treated with the AGPC solution.

6.2.1. Solid tissues

1. Finely mince tissue sample (approximately 0.5-1.0 g) using sterile forceps and scissors on a clean petri dish
2. Place in a 50 mL polypropylene tube and add 4 mL of Solution D (Disruption solution-see above). Mix well with a probe blender or instead use a hand held glass homogeniser. Keep on ice as much as possible during handling procedures
3. Add 1/10 initial volume (0.4 mL) of 2M sodium acetate, pH 4.2 and mix
4. Add equal volume (4 mL) of unbuffered, water-saturated phenol (pure) and mix
5. Add 1/5 volume (0.8 mL) of chloroform/isoamyl alcohol (49:1), mix vigorously for 10 - 20 s and leave on ice for approximately 20 min
6. Centrifuge at 10,000 g for 20 min. In a high speed centrifuge (e.g., Sorvall centrifuge with an SS34 rotor)

7. Remove upper aqueous phase to a clean centrifuge tube and precipitate for at least 2h at -20°C with 2.5 volumes (10 mL) of absolute ethanol (or at -70°C for 1 h). The RNA is then pelleted at 10,000 rpm for 10 min, as in step 6, washed with 70% ethanol and pelleted by centrifugation at 10,000 g for 5 min
8. Drain and again wash with 70% ethanol and pellet as above. Dry the RNA pellet well (leave tube standing inverted on a clean tissue in a laminar flow hood) and dissolve RNA in 2 mL of sterile distilled water
9. Determine the RNA concentration and purity by absorption at 260 nm and 280 nm. If purity is satisfactory (260/280 ratio of 1.5 or higher) adjust concentration to 1mg/mL and store at -20°C. If too dilute, reprecipitate the RNA (add sodium acetate, pH 4.2 to 0.1M and 2.5 volumes of absolute ethanol) and redissolve the pellet at 1.0 mg/mL.
10. Note: If there is a lot of protein remaining it may be necessary to purify the RNA further by protease digestion. If the virus RNA concentration in the sample is a high positive results can be obtained with very impure samples i.e. ratios as low as 1.3 therefore, it is best to test the RNA at this stage and continue to stage 9 if results are negative
11. Add 2 mL of 2x pronase buffer containing 1 mg/mLpronase to the RNA dissolved in 2 mL distilled water and incubate at 37°C for 1-2 h to digest the remaining protein
12. Add equal vol (4 mL) buffer-saturated phenol-chloroform (1:1), mix vigorously for 1-2 min. Centrifuge at 880-900 g for 10 min to separate the phases
13. Take the aqueous phase with a pipette and place in a fresh tube
14. Add 2.5 volumes (10 mL) of absolute ethanol and precipitate at -20°C as before and repeat stage 8

Note: To extract RNA from peripheral blood mononuclear cells (PBMCs) the above procedure to stage 8 is carried out

It is not necessary to pronase treat the RNA as it is normally of high purity after the first extraction. All the volumes can be reduced to 1/10 if 10 mL or less whole blood is used as starting material and the high speed centrifugation steps carried out in a microcentrifuge.

6.2.2. Preparation of PBMCs

1. Collect blood in EDTA or heparin and mix gently to prevent partial clotting
2. If more than 10 mL of blood are to processed centrifuge at 2500 rpm (1300 g) for 10 min at room temperature (18-20°C) and collect the buffy coat layer and resuspend the cells in Hank's buffered saline solution (HBSS) in a final volume of 20 mL in a 50 mL polypropylene tube. PBS can be substituted for HBSS if necessary
3. If 10 mL or less are being processed start at step 2

6.2.2.1. Procedure for 10 mL or less whole blood

1. Collect 5-10 mL of whole blood and make up to 20 mL with HBSS
2. Mix and carefully underlay the blood with 10 mL of Ficoll solution (Lymphoprep or Nycoprep) and centrifuge at 2000 rpm (800-900 x g) for 30 min at room temperature to pellet the red cells
3. Collect the clear band of PBMCs floating on top of the ficoll layer, transfer to a clean 50 mL tube and resuspend cells in about 8 mL HBSS. Make up to 40-45 mL with HBSS.
4. Centrifuge at 1500 rpm (500 g for 10 min)
5. Wash pelleted PBMC with HBSS (2 times) by resuspending in 8-10 mL HBSS and mixing with another 30-40 mL. HBSS and pellet cells by centrifugation as in step 5.
6. Count cells. Dilute 10 µL aliquot 1:10 with Trypan blue to check viability if it is required to store viable cells at -70°C or in liquid nitrogen for subsequent virus isolation

Note: if field virus isolation is to be attempted it is best to try at this stage. Mix PBMCs with susceptible cells at a ratio of 1:5 (10^6 : 5 x10^6). Vero or primary bovine kidney cells are usually used but transformed lymphocyte lines are much more susceptible to rinderpest infection and should be used if routine isolations are being carried out (e.g., Theileria-transformed bovine lymphocytes or transformed marmoset lymphoid cells (B95a). The PBMCs can also be stored frozen in viable form.

6.2.3. RNA extraction from PBMCs

Part or all of the cells can be used for RT/PCR diagnosis of rinderpest or PPR.
Cells from a minimum volume of 5 mL of whole blood should be used for this.

1. Resuspend washed cell pellet in 1.0 mL HBSS and transfer to clean sterile 1.5mL Eppendorf microtubes
2. Centrifuge briefly (20-30 s) to pellet cells
3. Remove supernatant and if necessary the cell pellets can be stored frozen at this stage
4. Add 0.4 mL of solution D to the pellets, resuspend by vigorous vortex mixing and proceed as for tissue samples using 1/10 volumes of each reagent as described above
5. There will be too little RNA in these samples to determine the concentration by absorbance at 260 nm and the pellet should be dissolved in 20-50 μL sterile double distilled or DEPC-treated water

6.2.4. Procedure for freezing PBMCs

If it is required to store viable PBMCs for future virus isolation they can be stored indefinitely under liquid nitrogen using the following procedure:

1. Resuspend pelleted cells at 2 x 10^7 cells/mL in FCS containing 10% DMSO (tissue culture grade)
2. Place 0.5 mL in labelled, precooled freezer vials and leave on ice for 10 min
3. Wrap the tubes in cotton wool and place at -20°C in a polystyrene box for 30 min
4. Transfer the box to -70°C for at least 24 h (can be left at -70°C for several months if necessary)
5. Transfer to liquid nitrogen (gas phase) for indefinite storage
6. Dissolve extracted tissue RNA pellet in RNase-free DEPC-treated water at 1mg/mL.
7. In the case of RNA derived from PBMCs use 1/4 to 1/10 of the RNA extracted from 10 mL of whole blood

6 3. Prepare RNA for reverse transcription

a. 5 µL	RNA solution (approximately = 5 :g)
b. 2 µL	Random hexanucleotide primers (50 ng/µl)
c. 3 µL	DEPC-treated water
10 µL	**Total**

1. Incubate for 5 min at 70°C to disrupt any RNA secondary structure and cool at room temp for 10 min to allow primer annealing.

2. Centrifuge briefly (10-20 s) before opening to ensure all of the sample is at the bottom of the tube.

6.4. Prepare RT buffer mix

		Final concentrations in reaction
a.	4 µL 5 x RT buffer	50 mM Tris-HCl pH 8, 3mM $MgCl_2$, 75 mM KCl
b.	2 µL BSA (acetylated, 1mg/mL)	0.1 mg/mL
c.	2 µL DTT (0.1M)	10 nM
d.	1 µL dNTPs (10mM each)	0.5 mM each dNTP
e.	1 µL Reverse Transcriptase	200 units
	10 µL Total	

Note: Make enough mix for one extra tube and add 10 µL of this mix to each RNA sample.

Mix gently and centrifuge (10-20 s) briefly to ensure all the reactants are at the bottom of the tube. Incubate at room temperature for 5 min and then at 37°C for 30-60 min.

If necessary the cDNA product can be stored at -20°C until used for PCR.

6.5. Procedure for PCR Amplification

Take 5 µL of the cDNA product and add:

a.	5.0 µL	10 x PCR buffer
b.	1.5 µL	50 mM $MgCl_2$
c.	35.0 µL	Sterile distilled water
d.	0.5 µL	Taq-polymerase (5 U/:l)
e.	1 µL	dNTPs 10 mM each)
f.	1 µL (10 pmol)	Primer 1
g.	1 µL (10 pmol)	Primer 2)

(Primer sets depend on virus to be identified)

The $MgCl_2$ is now normally supplied separately and it is usually at 50mM concentration, however, check the concentration and adjust volume added and water volume accordingly.

- To make the final reaction solution 1.5 mM, add 1.5 μL of 50 mM $MgCl_2$ to each tube

- If the concentration of $MgCl_2$ is 25 mM then add 3.0 μL and reduce the water volume to 33.5 μL

Some manufacturers also supply enzyme stabilisers that may be added to the reactions.

If required, add the appropriate amount of stabiliser and reduce the water volume by the same amount. The reagents can be combined to make a master mix if many samples are to be assayed. In this case, 45 μL of the mix is added to 5 μL of the copy DNA product. However, if each reaction requires different sets of primers, they cannot added to the master mix but have be added separately to each tube.

In this case, 43 μL of the master mix, minus primers, is added to each tube.
Mix, briefly (10-20 s) centrifuge the mixture and overlay with 50 μL mineral oil.

Note: overlaying with mineral oil is not required if the thermocycler has a heated lid to prevent evaporation.

Place tubes in PCR machine and select the appropriate program.

6.6. PCR procedure

Any make of thermocycler may be used but the results may vary depending on the machine. Different time settings can be tried to get the optimum results, however, the following cycles have been found to work satisfactorily on several different types of commercial thermocycler.

6.6.1. Programming the machine

Use the Hot start approach (see chapter 2).

Step 1: temp	94°C	time	2 min
Denaturation:	94°C	time	45 s
Annealing:	50°C	time	45 s
Synthesis:	72°C	time	1 min
Number of cycles	30		
Final extension:	72°C	time	5 min

6.7. Analysis of RT-PCR products

An aliquot of each reaction (usually 8 µL) is taken, mixed with one quarter volume of the 5 x gel loading buffer and loaded on a 1.5% agarose gel.

Suitable markers, either 123 or 100 base pair ladders should always be used.
The size of the expected products is determined by the distance between the two primers in the primary sequence of the nucleic acid being analysed.

In the case of the universal primer set 1 this is 429 nucleotides and 372 nucleotides in the case of the F-specific gene primers sets. The position of the correct sized product can be determined by reference to the positive controls in the reactions and confirmed by the DNA marker ladder. Non-specific DNA products are sometimes present, usually in negative samples or when the concentration of the target nucleic acid sequence is low and care must be taken in the interpretation of the results.

In such cases a second test to confirm the identity of the DNA product should be carried out e.g., by restriction digestion, by hybridisation probe analysis or by sequence analysis. These latter two techniques will be demonstrated later in the course.

6.8. Cloning of PCR products

Taq polymerase adds an overhanging A residue at the 3' end of each DNA strand.

To clone these products efficiently it is necessary to use a special "T" vector system where an extra T residue has been added to the 5' end of each DNA strand in the vector.

These are available commercially from several companies.

Note: PCR is a very sensitive technique and extreme caution should be taken to avoid cross-contamination of samples.

6.9. Precautions to avoid cross-contamination

One of the most important outcomes of inexperience in this technique that has dangerous consequences is the possibility of obtaining a false positive result by careless handling of the materials and reagents.

- Ideally, preparation of the RNA, setting up of the RT-PCR and gel analysis of the amplified DNA should each be done in separate rooms using separate sets of pipettes
- The tubes containing the amplified DNAs should never be opened in the same place in which the reactions are set up, since aerosols of DNA can be produced and contaminate the atmosphere
- DNA is very stable in the environment and minute amounts can be carried in micro droplets to contaminate a newly opened reaction tube and amplified using such a sensitive technique as the PCR
- The same pipettes should never be used to set up the RT-PCR and to sample the resulting product; simply changing to a new tip is not sufficient, separate pipettes must be used for each procedure
- To minimise the possibility of aerosol contamination special aerosol resistant tips (ART tips) should always be used for setting up the RT-PCR reactions, or alternatively, special positive displacement pipettes
- **Each analysis** should always include both negative (to check for reagent contamination) and positive (to check for enzyme or machine failure) controls
- Water should be used in place of RNA as the negative control

Note: An alternative suggested in some manuals is to use the RNA but to omit the reverse transcriptase to prevent copy DNA production, but this is not a good negative control since some batches of Taq polymerase are known to have reverse transcriptase activity and this can lead to a false positive result even though there is no contamination of the reagents. As a positive control we routinely use RNA from a virus that is not expected to be present in the sample e.g., human measles virus is used as a positive control for RPV/PPRV diagnostic samples.

To check that the RNA is of sufficiently good quality to be copied by the reverse transcriptase, a primer set that amplifies part of a messenger RNA for a highly conserved cellular gene (e.g., actin gene) can be added to the RT-PCR reactions to check that the RNA is not completely degraded. A sample where the RNA is degraded by poor handling and extraction will give a negative result, regardless of whether it comes from an infected or an uninfected animal. The actin primer set can be added in the same tube as the PPR F1/F2 or the RPV F3/F4 primer sets since the amplified DNA products are sufficiently different in size to be distinguished on a 1.5% agarose gel.

6.10. Optimising the reactions

Since PCR is completely dependent on the primers annealing to their template cDNA, and the kinetics of this process are critically dependent on the base composition and sequence of the primers being used, the optimal conditions of salt concentration and temperature for PCR will vary with different primer sequences.

Factors such as the Mg^{2+} concentration and the annealing temperature can greatly affect the efficiency of the PCR reaction and the production of non-specific DNA products.

The conditions given for the morbillivirus test have been optimised for the both the Mg^{2+} and the primer annealing temperature but these parameters should be optimised for each new PCR developed.

6.11. Interpretation of results

In most cases there is usually only one DNA band of the expected size and interpretation is simple. However, in some diagnostic samples there may only be a low concentration of virus-specific RNA and non-specific amplification products close to the expected size for the specific DNA fragments may be produced. If there is no other tissue sample available in such cases it is necessary to confirm the virus-specific origin of the DNA product, either by hybridisation analysis or by using a second nested primer set that will only amplify a virus-specific DNA fragment and not the non-specific one.To do this 1 µL of the PCR product is subjected to a second PCR amplification using the required nested primer set in a 50 µL PCR reaction as above. This usually gives a clear result, but it can also be further confirmed by hybridisation probe analysis.

For comparisons with other viruses in circulation it is recommended that the DNA product be sequenced; this is particularly important when a positive virus identification is made in a suspect case from a non-endemic area and the likely source of infection needs to be traced.

6.12. Primer Sets

Primer	Gene (Position)	Sequence
UP-N1	N(MVN: 789>813)	5'ACAAACCNAGGATTGCTGAAATGAT
UP-N2	N(MVN: 1002<1026)	5'CTGAA(T/C)TTGTTCTGAA(T/C)TGAGTTCT
UP-P1	P(RPN:)	5' ATGTTTATGATCACAGCGGT
UP-P2	P(RPN:)	5' ATTGGGTTGCACCACTTGTC
F1	PPR-F (777>801)	5'ATCACAGTGTTAAAGCCTGTAGAGG
F1A	PPR-F (802>821)	5'ATGCTCTGTCAGTGATAACC
F1B	PPR-F (760>784)	5'AGTACAAAAGATTGCTGATCACAGT
F2	PPR-F (1124<1148)	5'GAGACTGAGTTTGTGACCTACAAGC
F2A	PPR-F (1092<1110)	5'CTATGAACAGAGGGGACAAG
F2D	PPR-F (1183<1207)	5'GGGTCTCGAAGGCTAGGCCCGAATA
PPR-DIG	PPR-F (847>871)	5'GAGACTGAGTTTGTGACCTACAAGC
F3	RPV-F (818>842)	5'GGGACAGTGCTTCAGCCTATTAAGG
F3A	RPV-F (845>865)	5'GCTCTGAACGCTATTACTAAG
F3B	RPV-F (801>825)	5'AGTATAAGAGGCTGTTGGGGACAGT
F4	RPV-F (1165<1189)	5'CAGCCCTAGCTTCTGACCCACGATA
F4A	RPV-F (1059<1079)	5'CTGCTTGTCGTATTTCCTCAA
F4D	RPV-F (1227<1249)	5'TGGGTCTCTGAGGCTGGGTCCAAAT
RPV-DIG	RPV-F (888>912)	5'CCACTACCAGTAGGAGACACAGAAG
P1	FMDV (VP1>477-496)	5'CCTACCTCCTTCAACTACGG
P2	FMDV (PB2A>34-47 and PB2>1-6)	5'GAAGGGCCCAGGGTTGGACTC

All the primer sequences for FMDV, RPV, PRV, BVDV used by the Pirbright laboratory are given.

Primer	Gene (Position)	Sequence
B1	BVD	5'GTAGTCGTCAGTGGTTCG
B2	BVD	5'GCCATGTACAGCAGAGAT
BA1*	Actin	5'GAGAAGCTGTGCTACGTCGC
BA2	Actin	5'CCAGACAGCACTGTGTTGGC
CDF1	CDV-F (662>686)	5'AATTTAGTCCTCGAACCAATCAACC
CDF2	CDV-F (1009<1033)	5'TAACCCTAATCTCTGCCCAACTAAT
MVF1	MV-F (824>848)	5'AGAACTGTTTTGGAACCAATTAGAG
MVF2	MV-F (1171<1195)	5'GAGCCCGAGCTTCTGGCCGATTAAA

*: The alpha-actin gene primers work on all domestic species e.g., cattle, sheep, goat, pig.

6.12.1. EMBL access numbers

RPV F gene: Z30700; PPR F gene: Z37017; RPV N gene: X68311; RPV P gene: X68311; CDV F gene: M21849; MV F gene: X16565

6.13. Correct sizes DNA fragments amplified by combinations of primer sets

Primer set	Gene	Size of DNA
F1b/F3d	PPR-F	448 *
F1/F2	PPR-F	372 * Nested sets
F1/F2a	PPR-F	334 *
F1a/F2	PPR-F	347
F1a/F2a	PPR-F	309
F3b/F4d	RPV-F	448 *
F3/F4	RPV-F	372 * Nested sets
F3a/F4a	RPV-F	235 *
F3/F4a	RPV-F	261
F3/F4d	RPV-F	431
F3a/F4	RPV-F	344
F3a/F4a	RPV-F	234
F3b/F4a	RPV-F	278
F3a/F4d	RPV-F	404
CDF1/CDF2 CDV-F		372
MVF1/MVF2	MV-F	372
N1/N2	N	234
UPPF/UPPR P		429
P1/P2	FMDV-VP1	214
B1/B2	BVDV	205
BA1/BA2	Actin	275

6.14. References

1. Chomczynski, P. & Sacchi, N., 1987. Single-step method of RNA isolation by acid guanidinium thiocyanate-phenol-chloroform extraction. Anal. Biochem., 162, 156-159

2. Forsyth, M. & Barrett, T. 1995. Evaluation of the polymerase chain reaction for the detection and characterisation of rinderpest and peste des petits ruminants viruses for epidemiological studies. Virus Research, 39, 151-163

7. The detection of Mycoplasma, contagious bovine pleuropneumonia and contagious caprine pleuropneumonia, using a PCR assay[8]

The purpose of this assay is to routinely screen diagnostic samples. This is a rapid detection method and can differentiate between Mycoplasma, CBPP and CCPP. This assay is applied to organ and cell culture samples.

Primers MM450/MM451/F-REAP3/R-REAP4/MGSO/GPO-3

7.1 Primer sequences

MM450	5'- GTA TTT TCC TTT CTA ATT TG –3'
MM451	5'- AAA TCA AAT TAA TAA GTT TG –3'
F-REAP3	5'- GAA ACG AAA GAT AAT ACC GCA TGT AG –3'
R-REAP4	5'- CCA CTT GTG CGG GTC CCC GTC –3'
MGSO	5'- TGC ACC ATC TGT CAC TCT GTT AAC CTC –3'
GPO-3	5'- GGG AGC AAA CAG GAT TAG ATA CCC T –3'

Type	Primer set	Amplicon size
CBPP	MM450 + MM451	574 bp
CCPP	F-REAP3 + R-REAP4	786 bp
Cell culture	MGSO + GPO-3	280 bp

7.2. Accredited Personnel

All PCR laboratory personnel

7.2.1. Safety

Hazardous or potentially hazardous samples must be processed in the laminar flow or Class II Biohazard cabinets. Correct protective gear e.g., lab coats and gloves must be worn at all times during the handling of Mycoplasma samples.

8 Courtesy of Onderstepoort Veterinary Institute, South Africa.

7.3. Accompanying documents

	Relevant SOP no:
Primer information	
Articles	
Solution and reagent preparation	
Safety procedure document	Safety manual
Sample reception SOP	
Dissemination of results	

7.4. Background principle

DNA is extracted from sample material, and amplified by Taq DNA polymerase.

Three tests can be done using three sets of primers, that will detect:

1. Presence of contagious bovine pleuropneumonia (CBPP)
2. Presence of contagious caprine pleuropneumonia (CCPP)
3. Mycoplasma contamination in cell cultures

The amplified DNA is then visualized by size fractionation on an agarose gel and the results recorded by using a photo documentation system.

7.5. Preconditions

All laboratory personnel must be familiar with the extraction and handling of DNA, and all procedures needed to perform PCR.

7.6. Procedure

7.6.1. Equipment

Micropipettes: 10 µL; 100 µL; 1000 µL

Micropipette filter tips: 10 µL

100 µL

1000 µL

Microcentrifuge tubes (1.5 mL)

Heating block 65°C and 100°C (e.g., Grant QBT1)

Microcentrifuge (e.g., Jouan)

Bench top microcentrifuge (e.g., Jouan, Heraeus)

Vortex apparatus (e.g., Heidolph)

Biohazard flow cabinet (N/A 6 ft LF)

Laminar flow cabinets (e.g., Labotec)

Thermocycler (e.g., Perkin-Elmer 9700; 9600; 2400)

Microwave oven (e.g., Panasonic)

Gel apparatus (e.g., EC Electrophoretic)

Photo documentation system (e.g., GeneGenius)

Photo documentation system (e.g., Lumi-Imager)

Ice machine (e.g., Scotsman)

7.6.2. Material and Reagents

dNTPs 10 mM (e.g., Lab. Spec.)

Oligonucleotides –

MM450/MM451/F-REAP3/R-REAP4/MGSO/GPO-3

20 pmol/µl (e.g., Whitehead)

DNA polymerase:

500 U Promega Taq DNA polymerase

500 U FastStart Taq DNA polymerase

Chloroform (e.g., Merck)

Absolute ethanol (e.g., Merck)

70% Ethanol

Biophenol (e.g., Biosolve)

DEPC-H_2O

1 x TAE buffer

Agarose (e.g., Separations)

DNA molecular weight marker

Loading buffer

Nucleon BACC2 Extraction and Purification kit (e.g.,(Amersham Pharmacia)

Molecular Diagnostic Services (MDS) Extraction kit

7.6.3. Control Reference Material

Test controls:

7.6.3.1. Positive controls

- DNA from Mycoplasma CBPP, CCPP and cell culture positive samples
- A high titre positive sample.
- A low titre positive sample (or the above sample suitably diluted).

7.6.3.2. Negative controls

- DNA from Mycoplasma CBPP, CCPP and cell culture negative samples.
- DEPC-H_2O.

7.7. Performance of assay

7.7.1. Sample preparation

The correct protective gear e.g., lab coats and gloves must be worn at all times during sample preparation.

7.7.2. Organ and cell culture samples

Collect a copy of the registration form on the sample reception board.
Collect samples from 4°C refrigerator.
Proceed with the Amersham Nucleon BACC2 extraction method, MDS extraction method or Crude method.

7.7.3. DNA Isolation (Amersham extraction kit)

7.7.3.1. Organ samples

1. Homogenise approximately 0.25 g of the sample in 2.5 mL of Reagent A of Nucleon BACC2 extraction kit
2. Incubate at room temperature (18-25°C) for 4 min, rotary mixing the tube constantly
3. Centrifuge at 1300 g for 10 min
4. Decant supernatant
5. Add 350 µL of Solution B and vortex
6. Add 100 µL of sodium perchlorate solution
7. Invert the tube 25-30 times
8. Add 600 µL of chloroform and vortex
9. Add 100 µL of resin. Do not shake the tube
10. Centrifuge at 300g for 3 min
11. Collect the upper clear phase (approximately 450 µL) and transfer to a clean tube
12. Add 900 µL of cold absolute ethanol. Mix by inversion until the precipitate appears
13. Centrifuge at 13000 rpm for 5 min
14. Decant supernatant
15. Add 500 µL of 75% Ethanol and vortex
16. Centrifuge at 13000 rpm for 5 min
17. Decant supernatant
18. Dry the pellet for 20 min, ensuring that all the ethanol has been removed
19. Dissolve in 200 µL of HPLC water
20. Store at -20°C

7.7.4. DNA Isolation (MDS extraction kit)

1. Homogenise approximately 0.25 g of the sample in 1.2 mL of Reagent 1
2. Vortex for 10 s
3. Centrifuge at 13000 rpm for 3 min
4. Decant the supernatant
5. Add 400 µL of Reagent 2 to the pellet
6. Vortex for 10 s to resuspend the pellet
7. Add 50 µL of Reagent 3 and vortex for 10 s
8. Incubate at 65°C for 5 min
9. Add 200 µL of chloroform and vortex for 10 s
10. Centrifuge at 13000 rpm for 1 min
11. Transfer the aqueous upper phase to a clean tube
12. Add an equal volume of Biophenol to the sample and vortex for 10 s
13. Centrifuge at 13000 rpm for 1 min
14. Transfer the supernatant to a clean tube
15. Repeat the previous three step until the inter phase is clean of proteins
16. Add 1 mL of absolute ethanol and mix by inversion until the DNA is completely precipitated (visible as white threads)
17. Centrifuge at 13000 rpm for 2 min
18. Discard supernatant
19. Add 500 µL of 70 % ethanol and vortex
20. Centrifuge at 13000 rpm for 1 min
21. Discard supernatant
22. Air dry pellet for approximately 10-15 min
23. Dissolve the pellet in 50 µL of HPLC water
24. Store at -20°C

*7.7.5. **Organ and cell culture samples***

1. Use 200 µL of sample
2. Boil at 100°C for 20 min
3. Snap cool on ice for 5 min
4. Centrifuge at 13000 rpm for 5 min
5. Transfer the supernatant to a clean tube and store at -20°C

7.8. PCR Method

7.8.1. Master mix for Mycoplasma CBPP and CCPP (x 1)

1. 2.5 µL 10 x Dynazyme buffer (see solution and reagent preparations manual)
2. 0.5 µL dNTPs (10 mM)
3. 0.5 µL each of MM450 and MM451 (20 pmol/µL) primers for CBPP, F-REAP3 and R-REAP4 (20 pmol/µl) primers for CCPP
4. 17.75 µL H_2O
5. 0.25 µL or 2U Taq polymerase
6. 3 µL of isolated DNA

7.8.2. Master mix for Mycoplasma cell culture (x 1)

1. 2.5 µL 10 x PCR2 buffer (see solution and reagent preparations manual)
2. 0.5 µL dNTPs (10 mM)
3. 0.5 µL each of MGSO and GPO-3 (20 pmol/µL) primers
4. 17.75 µL H_2O
5. 0.25 µL or 2U Taq polymerase
6. 3 µL of isolated DNA

7.9. PCR amplification for Mycoplasma CBPP

Initial denaturation	95°C	5 m
Denaturation	94°C	30 s
Annealing	50°C	30 s
Extension	72°C	30 s
Number of cycles	30	
Elongation	72°C	5 m

7.10. PCR amplification for Mycoplasma CCPP

Initial denaturation	96°C	3 min
Denaturation	96°C	30 s
Annealing	60°C	30 s
Extension	72°C	60 s
Number of cycles	33	

7.11. PCR amplification for Mycoplasma cell culture

Initial denaturation	95°C	5 min
Denaturation	94°C	30 s
Annealing	58°C	30 s
Extension	72°C	30 s
Number of cycles	33	
Elongation	72°C	5 m

7.12. Gel separation

1. A 1.5 % agarose gel stained with ethidium bromide is used (Appendix B)
2. 10 µL of the product is loaded with 2 µL loading buffer
3. 2 µL of a 100 bp DNA molecular weight marker is loaded with 2 µl loading buffer in a single outside well
4. Gel electrophoresis is performed at 100 to 120 V for 30 min

Analysis is done by using an automated photo documentation system.

7.13. Analysis and evaluation

The positive control should yield a fragment size of

574 bp for Mycoplasma CBPP

786 bp for Mycoplasma CCPP

280 bp for Mycoplasma cell culture

The negative control should not show any amplicons.

The results are interpreted by accredited Personnel by looking at the case history and correlating the results with previous results.

Results are documented in the specific assay file. Record keeping is the responsibility of the laboratory manager.

A report is sent to the submitting veterinarian containing client information and test results obtained. An explanation of the results and/or any recommendations for future testing may be included where warranted. This report is printed out on a letter-head, signed by the laboratory manager, and an electronic or hard copy is stored.

The copy of the report is then faxed through to the veterinarian within 12 hours after results are obtained. The letter is also sent by mail the following day. A copy of the report is kept in the appropriate file. A computer-based software programme can be used to register arrival of samples at the section and also for the sending out of accounts to clients.

Evaluation and dissemination of results.

7.14. Variations

Inhibition of the PCR assay can occur on account of the following conditions:

Observation	Action
Inhibitors in the PCR reaction (protein)	Repeat the sample extraction
Dnase	Request new sample
Faulty primer	Order new primers
Inappropriate [Mg^{2+}] concentration	Repeat PCR with a higher [Mg^{2+}] concentration.

7.15. References

1. Bashiruddin, J.B., Taylor, T.K. and Gould, A.R. 1994. A PCR-based test for the identification of *Mycoplasma mycoides* subspecies *mycoides* SC. J Vet Diagn Invest., 6, 428-434.

2. Johansson, K.E., Persson, A. and Persson, M. 1997. Diagnosis of contagious caprine and contagious bovine pleuropneumonia by PCR and restriction enzyme analysis. Proceedings of the International Symposium on diagnosis and control of livestock diseases using nuclear and related techniques, Vienna.

3. Van Kuppeveld, F.J.M., Johansson, K.-E., Galama, J.M.D., Kissing, J., Bölske, G., Van der Logt, J.T.M. and Melchers, W.J.G. (1994). Detection of Mycoplasma contamination in cell cultures by a Mycoplasma group-specific PCR. Applied and Environmental Microbiology, 60 (1), 149-152.

8. The detection of Rift valley fever virus using a PCR assay[9]

This is a rapid detection method and will assist in the identification of Rift Valley fever virus infection. This assay is applied to blood and serum samples.

(Primers RVF1-F/RVF2-R)

8.1. Primer sequences

RVF1-F	5'- TGT CAC ACT GCT CTC AGT GCC –3'
RVF2-R	5'- GGA GCT TGC CTG AAT CAG TTC –3'

Type	Primer set	Amplicon size
Rift Valley fever virus	RVF1-F + RVF2-L	363 bp

8.2. Safety

Hazardous or potentially hazardous samples must be processed in the laminar flow or Class II Biohazard cabinets. Correct protective gear e.g., lab coats and gloves must be worn at all times during the handling of Rift valley fever samples (see Safety Manual).

8.3. Accompanying documents

Primer information	
Articles	
Solution and reagent preparation	
Working instructions manual SOP	
Safety procedure document	Safety manual
Sample reception SOP	
Dissemination of results	

[9] Coutesy of Onderstepoort Verinary institute, South Africa.

8.4. Background principle

Viral RNA is extracted from sample material, then converted into cDNA using reverse transcriptase and amplified thereafter by a Taq DNA polymerase (one-step RT-PCR).

The amplified DNA is then visualised by size fractionation on an agarose gel and the results recorded by using a photo documentation system, the Syngene GeneGenius Bio Imaging System.

8.5. Preconditions

All laboratory personnel must be familiar with the extraction and handling of RNA, and all procedures needed to perform RT-PCR.

8.6. Procedure

8.6.1 Equipment

Micropipettes 10 µL; 100 µL; 1000 µL
Micropipette filter tips: 10 µL
100 µL
1000 µL
Microcentrifuge tubes (1.5 mL)
Microcentrifuge (e.g., Jouan)
Bench top microcentrifuge (e.g., Jouan, Heraeus)
Vortex apparatus (e.g., Heidolph)
Biohazard flow cabinet (N/A 6 ft LF)
Laminar flow cabinets (e.g., Labotec)
Thermocycler (e.g., Perkin-Elmer 9700; 9600; 2400)
Microwave oven (e.g., Panasonic)
Gel apparatus (e.g., EC Electrophoretic)
Photo documentation system (e.g., GeneGenius)
Photo documentation system (e.g., Lumi-Imager)
Ice machine (e.g., Scotsman)

8.6.2. Material and Reagents

dNTPs 10 mM (Lab. Spec.)

Oligonucleotides – RVF1-F/RVF2-R 5 pmol/µL

DNA polymerase:

250 U HotStar Taq DNA polymerase

250 U TaKaRa Ex Taq

500 U FastStart Taq DNA polymerase

MMLV-RT 200 U/µl (Promega)

HPRI 110 U/µl (AEC-Amersham)

Isopropanol (e.g., Ass. Chem. Enterprises)

70% Ethanol

DEPC-H_2O

1 x TAE buffer

Agarose (e.g., Separations)

DNA molecular weight marker

Loading buffer

Biosolve (Sigma)

8.6.3. Control Reference Material

Test controls

8.6.3.1. Positive controls

- Serum from naïve sheep spiked with a known amount of virus-containing cell culture supernatant.
- A high titre positive sample.

8.6.3.2. Negative controls

- Naïve serum or blood from sheep.
- DEPC-H_2O

8.7. Performance of assay

8.7.1. Sample preparation

The correct protective gear e.g., lab coats and gloves must be worn at all times during sample preparation.

8.7.1.1. Serum sample

1. Collect copy of the registration form on the sample reception board
2. Collect samples from -70°C freezer
3. Leave at room temperature (18-25°C) to thaw
4. Collect 200 μL of the sample volume and aliquot into a separate microcentrifuge tube
5. Proceed with the Chomczynski one-step RNA extraction method

8.7.1.2. Blood sample

1. Collect copy of the registration form on the sample reception board
2. Collect samples from -70°C freezer
3. Leave at room temperature (18-25°C) to thaw
4. Use 200 μL from each sample
5. Proceed with the Chomczynski one-step RNA extraction method

8.8. RNA Isolation

8.8.1. Solution D / Chomczynski extraction

1. Collect 200 μL blood or serum in a 1.5 mL microcentrifuge tube
2. Add 500 μL solution D
3. Shake well to mix
4. Add 50 μL 2M NaAc
5. Shake well to mix
6. Add 500 μL Biosolve
7. Shake well to mix
8. Centrifuge for 15 min at 14000 rpm
9. Transfer the supernatant fluid to a microcentrifuge tube containing 600 μl 100% isopropanol.
10. Place at -20 °C for at least 20 m
11. Centrifuge for 20 min at 14000 rpm
12. Remove supernatant. Take care not to lose the pellet
13. Wash the pellet with 70 % ethanol
14. Centrifuge 14000 rpm for 10 m
15. Remove the supernatant. Take care not to lose the pellet
16. Air-dry the pellet
17. Dissolve the pellet in 20μL HPLC grade water
18. Dilute RNA 1:10 before use in RT-PCR reaction
19. Store at -20 °C

8.9. Preparation of solutions

8.9.1.1. 5 M Guanidium thiocyanate

23.632g Guanidium thiocyanate (MW=118.16) dissolved into 40 mL water.

8.9.1.2. 1 M Sodium citrate

1. To prepare 1M tri-sodium citrate.2 hydrate, add 14.705g (MW=294.1) to 50 mL water
2. To prepare 1M sodium citric acid (MW=210.14), add 10.507g to 50 mL water.
3. Determine the pH of both solutions
4. Take 20 mL of the solution with the pH closest to 7 and add the other solution until the pH is 7. Store at -20 °C

8.9.1.3. 10 % Sarcosyl

1. 2.0 g N.lauroyl-sarcosine sodium salt (MW=293.4) added to 20 mL water
2. For 2 M Sodium acetate add 32.812g of sodium acetate (MW=82.03g) to 20 mL of water. Adjust the pH to 4 using acetic acid
3. Store at -20 °C

8.9.1.4. Solution D

1.	5 M guanidium thiocyanate	40 mL
2.	1 M sodium citrate (pH 7)	1.25 mL
3.	10% Sarcosyl	1.25 mL
4.	2-mercaptoethanol	0.36 mL
5.	HPLC grade water	7.14 mL
6.	Total	50 mL

8.10. RT-PCR Method

8.10.1. Master mix (x 1)

1. 2.5 µL 10 x HotStar/Takara Ex Taq/FastStart buffer
2. 1 µL dNTPs (10 mM)
3. 1 µL each of RVF1-F and RVF2-R (5 pmol/µl) primers
4. 0.1 µL HPRI (110 µl)
5. 0.1 µL MMLV-RT (200 u/µl)
6. 13.8 µL H_2O
7. 0.5 µL Taq polymerase (HotStar Taq, TaKaRa Ex Taq or FastStart Taq).
8. 5 µL of isolated viral RNA (diluted 1:10)

8.10.2. cDNA synthesis

Reaction is carried out in a Perkin Elmer GeneAmp PCR 9700/9600/2400 system at 37°C for 30 min.

The reaction is immediately placed on ice and stored at -20°C.

PCR amplification:

Denaturation	94°C	30 s
Annealing	65°C	30 s
Extension	72°C	30 s
Number of cycles	35	
Elongation	72°C	7 min

8.10.3. Gel separation

- A 2 % agarose gel stained with ethidium bromide is used
- 10 µL of the product is loaded with 2 µL loading buffer
- 2 µL of a 100 bp DNA molecular weight marker is loaded with 2 µL loading buffer in a single outside well
- Gel electrophoresis is performed at 100 to 120 V for 30 min
- Analysis is done by using an automated photo documentation system

8.11. Analysis and evaluation

The positive control should yield a fragment size of 363 bp. The negative control should not show any amplicons

The results are interpreted by also looking at the case history, correlating the results with previous results.

Results are documented in the specific assay file. A report is sent to the submitting veterinarian containing client information and test results obtained. An explanation of the results and/or any recommendations for future testing may be included where warranted.

This report is printed out on a letterhead, signed by the Division head or his deputy, and an electronic copy or hard copy is stored. A copy of the report is then faxed through to the veterinarian within 12 h after results are obtained. The letter is also sent by mail the following day.

8.12.1 Evaluation and dissemination of results

Analytical sensitivity = 0.25 $TCID_{50}$

The sensitivity was determined by a 10-fold serial dilution of virulent virus-containing cell culture supernatant with a titre of 10^6 $TCID_{50}$/mL into naïve serum. A 5 µL aliquot of 20 µL RNA extracted from 200 µL of serum was used as template.

8.12. Variations

Variation of the RT-PCR assay can occur on account of the following conditions:

Observation	Action
Inhibitors in the RT-PCR reaction (protein)	Repeat the sample extraction
Rnase	Request new sample
Faulty primer design	Order new primers
Inappropriate [Mg^{2+}] concentration	Repeat the PCR with a higher/lower [Mg^{2+}] concentration.

8.13. References

1. Collett, M.S., Purchio, A.F., Keegan, K., Frazier, S., Hays, W., Anderson, D.K., Parker, M.D., Schmaljohn, C., Schmidt, J. and Dalrymple, J.M. 1985. Complete nucleotide sequence of the m-RNA segment of Rift Valley Fever Virus. Virology, 144, 228-245

9. The Detection of Lumpy skin disease virus using a PCR assay[10]

The purpose of this assay is to routinely screen diagnostic samples for lumpy skin disease virus infection (LSDV). This assay is applied to whole blood and organ samples

(Primers OP3 and OP49)

9.1. Primer sequences

OP3	5'-CACCAGAGCCGATAAC–3'
OP49	5'-GTGCTATCTAGTGCAGCTAT–3'

Type	Primer set	Amplicon size
Lumpy Skin Disease Virus	OP3 + OP49	450 bp

All PCR personnel are assigned to the lumpy skin disease virus PCR

9.2. Safety

Hazardous or potentially hazardous samples must be processed in the laminar flow or Class II Biohazard cabinets. Correct protective gear e.g., lab coats and gloves must be worn at all times during the handling of Lumpy skin disease virus samples.

9.3. Accompanying documents

Primer information	
Articles	
Solution and reagent preparation	
Working instructions manual SOP	
Safety procedure document	Safety manual
Sample reception SOP	
Dissemination of results	

[10] Coutesy of Onderstepoort Verinary institute, South Africa.

9.4. Principle of the assay

DNA is extracted from sample material, and LSDV-specific DNA is amplified by Taq DNA polymerase. The amplified DNA is then visualised by size fractionation on an agarose gel and the results recorded by using a photo documentation system.

9.5. Preconditions

All laboratory personnel must be familiar with the extraction and handling of DNA, and all procedures needed to perform PCR.

9.6. Procedure

9.6.1. Equipment

Micropipettes 10 μL; 100 μL; 1000 μL (e.g., Eppendorf)

Micropipette filter tips: 10 μL
100 μL
1000 μL
Microcentrifuge tubes (1.5 mL)
Microcentrifuge (e.g., Jouan)
Bench top microcentrifuge (e.g., Jouan, Heraeus)
Vortex apparatus (e.g., Heidolph)
Biohazard flow cabinet (N/A 6 ft LF)
Laminar flow cabinets (e.g., Labotec)
Thermocycler (e.g., Perkin-Elmer 9700; 9600; 2400)
Microwave oven (e.g., Panasonic)
Gel apparatus (e.g., EC Electrophoretic)
Photo documentation system (e.g., GeneGenius)
Photo documentation system (e.g., Lumi-Imager)
Ice machine (e.g., Scotsman)
Water bath
Waterbath sonicator

9.6.2. Materials and Reagents

dNTPs 10 mM (Lab. Spec.)
Oligonucleotides – OP3/OP49
20 pmol/µL (e.g., Lab Spec Services) DNA polymerase:
250 U Takara Ex Taq DNA polymerase
SDS (BDH)
Proteinase K (e.g., Sigma)
Absolute ethanol (e.g., Merck)
tRNA (Roche)
PBS
Phenol:Chloroform:Isoamylalcohol (e.g., Sigma)
HPLC Sterile Water (e.g., Sigma)
B-mercaptoethanol (e.g., Merck)
EDTA (ICN)
NaAc (Merck)
DEPC-H_2O
1 x TAE buffer
Agarose (e.g., Separations)
DNA molecular weight marker
Loading buffer

9.6.3. Control Reference Material

Test controls:

9.6.3.1. Positive controls

- DNA from LSDV-positive field samples
- A high titre positive sample.
- A low titre positive sample (or the above sample suitably diluted).

9.6.3.2. Negative controls

- DNA from LSDV-negative field samples. DEPC-H_2O

9.7. Performance of assay

9.7.1. Sample preparation

The correct protective gear e.g., lab coats and gloves must be worn at all times during sample preparation.

Collect copy of the registration form on the sample reception board.
Collect samples from 4°C fridge.

9.8. DNA Isolation

9.8.1. Whole blood sample

1. Take 1mL of whole blood and sonicate in a waterbath sonicator for 15 min
2. Centrifuge at low speed (1000 rpm) for 5 min
3. Transfer the supernatant to a clean microcentrifuge tube
4. Centrifuge at high speed (13000 rpm) for 20 min
5. Discard the supernatant
6. Resuspend pellet in 100 µL PBS
7. Add 100 µL lysis buffer (1% SDS, 20 mM β–mercaptoethanol, 20 mM EDTA)
8. Digest with Proteinase K (final concentration, 100 µg/mL) at 56°C for 2 h
9. Add one volume Phenol:Chloroform:Isoamylalcohol (25:24:1)
10. Mix well by inversion
11. Centrifuge at 13000 rpm for 5 min
12. Transfer the upper aqueous layer to a clean microcentrifuge tube
13. Add 2.5 volumes absolute ethanol, 0.5 µL of 10 mg/mL calf liver tRNA and 1/10th volume of 5 M sodium acetate (pH 5.3). Mix thoroughly
14. Precipitate overnight at –20°C
15. Pellet the DNA by centrifugation at high speed (13000 rpm) for 15 min
16. Discard the supernatant
17. Air-dry the pellet
18. Resuspend the DNA pellet in 50 µL sterile, ultra-pure water
19. Store at -20°C

9.8.2. Skin lesions

1. Obtain lesion in native state from animal, i.e., do not store in any liquid medium, or fix etc
2. Store at 4°C or –20°C until required for further analysis
3. Cut a thin section through the lesion in a sterile manner and chop section into small pieces
4. Macerate tissue using two 19 gauge needles or a homogeniser
5. Add 1 mL PBS
6. Mix well, and allow large pieces to settle for 3 min
7. Transfer the supernatant to a clean microcentrifuge tube
8. Centrifuge at 1000 rpm for 5 min
9. Discard the supernatant and resuspend pellet in 200 µL PBS
10. Sonicate in a waterbath sonicator for 15 min
11. Divide into two equal aliquots, freeze one stored at–20°C for future reference.
12. Add 100 µL lysis buffer to the other aliquot
13. Digest with Proteinase K (final concentration, 100 µg/mL) at 56 °C for 2 h
14. Add one volume Phenol:Chloroform:Isoamylalcohol (25:24:1)
15. Mix well by inversion
16. Centrifuge at 13000 rpm for 5 min
17. Transfer the upper aqueous layer to a clean microcentrifuge tube
18. Add 2.5 volumes absolute ethanol, 0.5 µl of 10 mg/mL calf liver tRNA and 1/10th volume of 5 M sodium acetate (pH 5.3). Mix thoroughly
19. Precipitate overnight at –20 °C
20. Pellet the DNA by centrifugation at high speed (13000 rpm) for 15 min
21. Discard the supernatant
22. Air-dry the pellet
23. Resuspend the DNA pellet in 50 µL sterile, ultra-pure water
24. Store at –20 °C

9.9. PCR Method

9.9.1. Master mix (x 1)

1. 2.5 µL 10 x Dynazyme buffer
2. 0.5 µL dNTPs (10 mM)
3. 1 µL each of OP3 and OP49 (20 pmol/µl) primers
4. 14.8 µL H_2O
5. 0.20 µL or 1u Takara Ex Taq polymerase
6. 5 µL of isolated DNA

9.10. PCR amplification

Denaturation	95°C	45 s
Annealing	53°C	45 s
Extension	72°C	60 s
Number of cycles	35	

9.10.1. Gel separation

A 1.0 % agarose gel stained with ethidium bromide is used

- 10 µL of the product is loaded with 1 µl 10 x loading buffer
- 2 µL of a 100 bp DNA molecular weight marker is loaded with 2 µL loading buffer in a single outside well
- Gel electrophoresis is performed at 70 to 80 V for 60 min
- Analysis is done by using an automated photo documentation system

9.10.2. Analysis and evaluation

The positive control should yield a fragment size of 450 bp.
The negative control should not show any amplicons.
The results are interpreted by looking also at the case history and correlating the results with previous results.
Results are documented in the specific assay file.
A report is sent to the submitting veterinarian containing client information and test results obtained. An explanation of the results and/or any recommendations for future testing may be included where warranted.

This report is printed out on a letterhead, signed by the Division head or his deputy, and an electronic copy or hard copy is stored. A copy of the report is then faxed through to the veterinarian within 12 hours after results are obtained. The letter is also sent by mail the following day.

9.11. Variations

Variations in the PCR assay can occur on account of the following conditions:

Observation	**Action**
Inhibitors in the PCR reaction (protein)	Repeat the sample extraction
Dnase	Request new sample
Faulty primer	Order new primers
Inappropriate [Mg^{2+}] concentration	Repeat the PCR with a higher/lower [Mg^{2+}] concentration

9.12. References

Weiss, K..E. 1968. Lumpy skin disease. Virology Monographs, 3, 112-282

APPENDICES

TECHNICAL INFORMATION AND GENERAL PROTOCOLS

APPENDIX A

THERMAL CYCLERS

Thermal cyclers / thermocyclers are microprocessor controlled machines that have allowed for the automation of PCR. Modern thermocyclers possess user-defined, temperature-controlled sample blocks or wells that provide both uniform temperature adjustment during cycling and high cycling time reproducibility. Most thermal cyclers use thermal engines based on the Peltier effect, in which heat is transferred from one side of a semiconductor to another, eliminating the need for refrigerants and compressors. Other systems use circulating air or water baths, or a combination of Peltier, resistive, or convective technologies. Either metal blocks, water, air or microwaves have been used for performing temperature transitions.

After the first thermocycler introduced by Perkin_Elmer Cetus, Emeryville, CA, numerous devices followed with various modifications (Table 1). Metal block PCR machines suffer from inconstant heat transfer from the heating block to the vessels if the contact between the two materials is not very tight, as well as the slowness and high weight of the devices if the performance of the peltier elements is not sufficient.

Where water baths with fluidic switching or mechanical transfer is used, the danger of contamination through contact with the water, as well as different temperatures in the water whenever the temperature controller and the location and extent of water inflow are not ideally matched.

An advantage is that temperature transfer into the tuber is limited only by heat conduction properties of the microfuge tube, while the response times of metal blocks are also limited by their heat capacity. Transition times are also lengthy.

Thin-walled reaction tubes also help to reduce reaction times. The use of glass or plastic capillaries also reduces reaction times, allowing for rapid heat transferral and also uniform temperature transition. When air is used for heat transferral, even shorter amplification times can be used **(Table 1).**

1. Peltier devices

Peltier devices, also known as thermoelectric (TE) modules, are small solid-state devices that function as heat pumps. A "typical" unit is a few millimeters thick by a few millimeters to a few centimeters square. It is a sandwich formed by two ceramic plates with an array of small bismuth telluride cubes ("couples") in between. When a DC current is applied, heat is moved from one side of the device to the other, where it must be removed with a heatsink. The "cold" side is commonly used to cool an electronic device such as a microprocessor or a photodetector. If the current is reversed the device makes an excellent heater. They are not very "efficient" and require a great deal of power. However, this disadvantage is more than offset by the advantage, that include: absence of moving parts, Freon refrigerant, noise and vibration, very small size, long life, capability of precision temperature control, etc.

1.1. Rate of Temperature Change

Peltier device cooling & heating speed: These devices can change temperature extremely quickly, but to avoid damage from thermal expansion control the rate of change is controlled to about 1 degree Celsius per second.

1.2. Heat and Cool

As discussed above, peltier devices make great heaters (when the polarity of the power supply is reversed) and thus their application and role in PCR. These devices are not all equal and have specific temperature ratings, usually 80°C for standard models to 200°C for high temperature models.

1.3. Ramp Times

The ramp time is defined as the time needed to go from one temperature to the next for the cycler block. When the ramp time is set to the minimum ramp time of zero, the rate of temperature change in the thermal cycler block is approximately 1°C/s for lamp heated blocks and 0.2°C/s for peltier heated blocks. For example, the shortest possible time necessary to ramp from 50°C to 95°C is approximately 45 s. The maximum ramp time is 9 min and 59 s.

Table 1. Comparison of some thermocyclers

COMPANY	Product	W xD xH (cm)	Temp Control	PCR Licence Rights	Programs Steps/ cycles	Heat Rate °C /sec	Cool Rate °C /sec	Thermal Range (°C)	Base Price $
Bio-Rad 800-424-6123 www.bio-rad.com	Gene Cycler'	27x19x16	Peltier	Yes	100/5/300	1.5	0.8	Amb-97	Call
	iCycler'	10x21x9	Peltier	Yes	255	3.0	2.0	4-10510.3	
Cepheid 408-541-4191 www.cepheid.com	Smart cycler-	30x3Ox22	LED/fan	OEM	150 programs	7.0	2.0	Call	IOEM
Eppendorf Scientific Inc. 800-421-9988 www.eppendorfsi.com	Master cycler®	26x41x27	Peltier	Yes	100/99/99	3.0	2.0	4-99 10.6	Call
	Master cycler gradient	26x41x27	Peltier	Yes	100/99/99	3.0	2.0	4-99 10.6	
	Master cycler personal	18x35x25	Peltier	Yes	100/99199	3.0	2.0	4-9910.6	
Ericomp 800-541-8471 www.ericomp.com	Delta Cycler I System	32x33x19	Peltier	No	26/99/99	1.3	1.3	4-11010.5	4,595
	Delta/ Cycler II System	32x44x19	Peltier	No	26/99/99	1.3	1.3	4-11010.5	7,595
	Power Block I System	32x33x19	Peltier	No	26/99/99	1.6	1.6	4-110 104	2,895
	Power Block II System	32x33x19	Peltier	No	26/99/99	1.6	1.6	4-11010.4	4,695
	Single Block System	36x41x15	Resistive/ water	No	26/99/99	1.3	1.3	Amb-11010.5	3,395
	Twin Block System	36x51x15	Resistive/ water	No	26/99/99	1.3	1.3	Amb-11010.5	5.595
Genex 44 1731 556617 www.genex.ltd.uk	Pacer	Call	Peltier	No	Unlimited	2.0	2.0	Call	Call

Table 1 (continued) Comparison of some thermocyclers

GL Applied Research Inc. 847-223-2220 members.aol.com/ glapplied/ homepage.htm	GTC-2 Genetic Thermal Cycler	46x6 3x48	Resistive/ water	No	99 programs	1.0	1.0	0-99 10.5	Call
Hybaid 888-449-2243 www.hybaid-us.com	Multi Block	Call	Peltier	Yes	>100/10/9 9	3.0	3.0	4-9910.4	Call
	PCR Express	23x3 9x25	Peltier	Yes	99/10/99	Up to 3.0	Up to 2.0	4-9910.4	5,800-
									6,310
	PCR Sprint	18x3 0x23	Peltier	Yes	60/5/99	Up to 3.0	Up to 2.0	4-9910.5	2,495
	Omn-E	19x2 8x19	Electric/fa n	Yes	36/10/99	Call	Call	Amb a10 to 99	1,995-
									2,495
Idaho Technology 800-735-6544 www.idahotec.com	Rapid Cycler	30x2 3x37	Light	No	99/-199	10.0	>5.0	Call	Call
Intelligent Bio-Instruments 617-354-3830 www.intelligentbio.com	TC 1600	Call	Electric	No	Call	1.5	1.3	30-100	Call
MU Research 888-735-8437 www.mjr.com	Tetrad	37x5 5x25	Peltier-Joule	No	400 programs	3.0	2.5	-5-105 10.4	Call
	DNA Engine Series	24x3 3x24	Peltier-Joule	No	200 programs	3.0	Call	-5-105 10.4	
	Mini Cycler	22x2 8x11	Peltier-Joule	No	80 programs	2.4	Call	-9-105 10.4	
	PTC-100	24x2 8x27	Peltier	No	340 programs	2.5	Call	0-100 10.4	
PE Biosystems 800-345-5224 www.pebio.com	DNA Thermal Cycler 2700	33x5 2x30	Cartridge	Yes	Call	1.0	1.0	-5-100 10.5	Call
	GeneAmp 2400	34x6 0x21	Cartridge	Yes		>1.0	>1.0	4-9910.5	
	GeneAmp 9600	Call	Film heater	Yes		1.0	1.0	4-9910.5	
	GerxeAm p 9700	Call	Peltier	Yes		1.6	1.6	4-99 10.5	

Table 1 (continued) Comparison of thermocyclers

Roche Molecular Biochemicals 800-428-5433 biochem.roche.com	Light Cycler	30x45x4O	Electric coil/au	Yes	Unlimited	20.0	20.0	Amb-98	55,00(
Snark Technologies 800-966-7808 www.snarktech.com	Snark 64 and 96 models	25x30x25	Peltier	No	350/no limit	4.4	3.2	0-105 10.5	3,495
St. John Associates 800-837-2023 www.stjohnassociates.com	BioTherm BioOven III	38x43x46	Circulating air	No	99/20/20	1.0	1.0	Amb-100	3,990
Stratagene 800-424-5444 www.stratagene.com	Robo Cycler® 40 & 96	31x49x21	Robotic hot block	Yes	99/16/99	No ramp'g	No ramp'g	25-99	6,995
	Infinity Temp Cycler	31x49x21	Robotic hot block	Yes	99/16/99	No ramp'g	No ramp'g	25-99	5,495
	Gradient Temp Cycler	31 x49x21	Robotic hot block	Yes	99/16/99	No ramp'g	No ramp'g	25-99	6,995
Tectne 800-225-9243 www.lechneuk.co.uk	Genius	22x42x24	Peltier	Call	99/9/99	2.8	1.8	4-9910.5	Call
	Progene	16x35x17	Peltier		50/9/99	2.8	2.2	4-9910.3	
	Touch Gene	22x42x26	Peltier		3/3/-	2.6	1.6	4-9910.5	
Whatman Biometra® 800-932-7250 www.biometra.com	Personal Cycler 20 & 48	Call	Peltier	No	100/99/99	2.0	>1.5	Call	Call
	3 Thermocycler		Peltier	No	Call	>1.5	>1.3		
	TGradient		Peltier	No	100 per directory	4.0	>3.0		
	TRIO Thermo block		Peltier	No	30/12199	1.5	1.3		
	UNO 11 Thermo cycler		Peltier	No	Call	3.3	2.4		
	T1 Thermo cycler		Peltier	No (	Call	4.0	3.0		

APPENDIX B

BASIC PROTOCOL FOR AGAROSE GEL ELECTROPHORESIS

1.1. Materials

(Buffer and reagent compositions are shown in Appendix J)

- Horizontal gel electrophoresis apparatus
- Microwave oven or heating plate
- Power supply
- UV transilluminator
- Electrophoresis buffer (1x TAE)
- Ethidium bromide (EtBr)
- Agarose
- Gel loading buffer (0.25 % bromophenol blue in 40 % sucrose)
- DNA molecular weight marker

1.2. Methods

1. Prepare an adequate volume of 1x TAE-buffer to fill the electrophoresis tank and prepare the gel. To facilitate visualization of DNA fragments during the run, ethidium bromide solution can be added to the electrophoresis buffer to a final concentration of 0.5 μg/mL
2. The amount of agarose that is used depends on the size of the tray and the concentration required. For general purposes a 1 % gel is used. Gels are typically between 0.5 and 1 cm thick

TRAY	1 x TAE BUFFER	1 % AGAROSE	1.5 % AGAROSE
7 x 10 cm	35 mL	0.35 g	0.52 g
15 x 10 cm	70 mL	0.70 g	1.05 g
15 x 15 cm	105 mL	1.05 g	1.57 g

3. Prepare a solution of molten agarose as required by adding the 1x TAE buffer to the appropriate amount of agarose powder in a suitable EM flask. Bring to boil in the microwave oven. The agarose must be completely dissolved prior to pouring. Adjust volume with distilled water to compensate for any loss.
4. Seal the ends of the plastic tray with masking tape and pour the gel after the agarose has been allowed to cool down to about 55°C - this prevents warping of the gel apparatus. Insert the comb and make sure that there are no bubbles trapped underneath the combs and that all bubbles on the surface of the agarose are removed before the gel sets.
5. After the gel has set, remove the tape from the casting tray and withdraw the gel comb, taking care not to tear the sample wells.
6. Place the gel casting tray containing the set gel in the electrophoresis tank. Add sufficient 1x TAE buffer to cover the gel to a depth of 1mm (or until the wells are just submerged). Make sure no air pockets are trapped within the wells.
7. Apply 5 µL of the gel loading buffer to each 10 µL sample and add samples to the individual wells. Take care not to overload and be sure to include appropriate DNA molecular weight markers.
8. Cover with the safety cover and run gel at 110 mA (7 x 10 cm tray) or 200 mA (15 x 10 cm tray) - typically 1 to 10 V/cm of gel. When the bromophenol blue marker is about two thirds from the top, the gel can be stained with ethidium bromide (EtBr).
9. Turn off the power supply when the bromophenol blue has migrated a distance judged sufficient for separation of the DNA fragments.
10. The DNA can be visualized on a UV transilluminator and photographed.
11. Gels that have been run in the absence of ethidium bromide can be stained by covering the gel in a dilute solution of ethidium bromide (0.5 µg/mL in water) and gently agitating for 10 to 30 min. If necessary, gels can be destained by shaking in water for an additional 30 min. This serves to remove excess EtBr that causes background fluorescence and makes visualization of small quantities of DNA difficult.

APPENDIX C

USING NUCLEIC ACID HYBRIDISATION IN ORDER TO VERIFY THE INTEGRITY OF A PCR AMPLICON

Various southern blot based techniques based on the transfer of the PCR amplicon to a nylon membrane and its visualisation with isotope or digoxigenin labelled (DIG) probes can be used. Only the DIG procedure will be discussed here due to its ease of use. The method uses DIG (a steroid hapten) to label DNA probes for hybridisation and subsequent color detection by enzyme immunoassay.

There are three steps involved in this procedure:

- **DNA labelling**

  DIG-labelled DNA probes are generated according to the random primed labelling technique.

- **Hybridisation**

  DIG-labelled probes are used for hybridisation to membrane-blotted nucleic acids according to standard methods. The use of the alkali-labile form of DIG-11-dUTP enables easier and more efficient stripping of blots for rehybridisation with a second labelled probe.

- **Immunological detection**

  The hybridised probes are immunodetected with anti-digoxigenin-AP, Fab fragments and are then visualized with the colorimetric substrates NBT/BCIP.

1. Preparation of DIG-labelled DNA probe (Roche; non-radioactive labelling and detection kit)

DIG -labelled DNA probes are generated according to the method of random primed labelling which is based on the hybridisation of random oligonucleotides to the denatured DNA template. The complementary DNA strand is synthesized by Klenow enzyme that uses the [3'OH] termini of the random oligonucleotides as primers and a mixture of deoxyribonucleotides containing DIG-11-dUTP, alkali-labile for elongation. DIG dUTP is incorporated every 20-25 nucleotides into the newly synthesized DNA. This density of DIG haptens in the DNA yields the highest sensitivity in the detection reaction.

1. If the DNA to be used as a probe is circular (e.g., a plasmid), it must first be linearised and then purified by phenol/chloroform extraction and ethanol precipitation

2. Denature the linearised DNA by boiling in a water bath/heating in a heating block for 10 min at 100°C and then immediately chill on ice

3. Pipette the following into a sterile Eppendorf tube:
   - 10 ng to 3 μg of the denatured, linear DNA
   - 2 μL Hexanucleotide mix
   - 2 μL dNTP-mix (dATP, dTTP, dCTP, dGTP, dIG-UTP)
   - Adjust the volume to 19 μL with sterile H_2O
   - 1 μL Klenow polymerase

4. Centrifuge the tube in a microfuge for a few seconds to mix the contents and incubate at 37°C for 1 to 20 h. (Longer incubations will increase the yield of labelled DNA)

5. Terminate the reaction by adding 2 μL of a 0.2 M EDTA solution (pH 8.0) to the reaction mixture

6. Precipitate the labelled DNA with 2.5 μL of a 4 M LiCl solution and 75μL ice-cold ethanol. Incubate for 10 min at -70°C or 2 h at -20°C

7. Collect the DNA by centrifugation at 15000 rpm for 15 min, wash with 70% ethanol and dry the DNA under vacuum. Resuspend the DNA in 25 μL of 1 × TE buffer

2. Southern Blot Technique

1. Sequentially soak the gel with gentle agitation in the following solutions
   (a) 200 mL Depurination solution (0.2 M HCl) for 15 min
   (b) Rinse 2 to 3 times with dH_2O
   (c) 200 mL denaturation solution (0.5 M NaOH, 1.5 M NaCl) for 30 min
   (d) 200 mL neutralization solution (1 M Tris.HCl, 1.5 M NaCl) for 30 min
2. While the gel is soaking, prepare the blotting set-up as follows:
   Cut 10 pieces of filter paper, 1 piece of nylon membrane and an 8-10 cm stack of paper towels to the size of the gel
3. Wet the membrane in a dish of dH_2O
4. Wet 6 pieces of filter paper with 20 × SSC and place on a piece of Gladwrap
5. Place the agarose gel onto the filters, smooth the gel and remove trapped air bubbles by rolling a glass pipette over the surface
6. Fold the Gladwrap tightly against the sides of the agarose gel to seal the paper/gel-sandwich properly. Do not cover any part of the top of the agarose gel with Gladwrap
7. Pre-wet the nylon membrane (same size as the gel) in 2 × SSC
   Carefully lay the membrane on top of the gel. Try to avoid trapping air bubbles under the membrane
   DNA transfer begins immediately therefore, do not move the membrane once it has touched the gel
8. Stack the remaining 4 sheets of filter paper on top of the membrane
   Wet 2 sheets with just enough 2 × SSC to saturate the paper
   Smooth out any air bubbles between the paper sheets and the membrane
9. Stack the pre-cut paper towels on top of the filter paper. It is important that the membrane, filter paper, or paper towels do not hang over the gel because they will bypass the movement of fluid through the gel
10. Lay a glass plate on top of the stack. Use a bottle containing approximately 500 mL of water as a weight by placing it on top of the glass plate
11. Leave at room temperature overnight for transfer to occur
    The rate of transfer of DNA to the membrane depends on the size of DNA fragment. Small fragments (about 1 kb) transfer within 1-2 h, but larger fragments (over 15 kb) require much longer
12. After transfer, remove the weight, towels, and filter paper above the gel
    Carefully remove the membrane and soak the membrane in 2 × SSC for 5 min
13. Fix the DNA onto the membrane by exposing the membrane to UV-illumination for 5 min on each side. Lay the membrane on a clean piece of filter paper and allow it to air-dry
14. Re-stain the agarose gel with EtBr and examine on the UV transilluminator to verify efficient DNA transfer

3. Hybridisation Step

1) Denature the DIG-labelled DNA probe (about 25 ng/mL) by boiling for 5 min and rapidly cooling in ice/water
2) Add the denatured probe to the pre-heated membrane (25°C) and mix well
3) Incubate for 6 h to over night with gentle agitation at 25°C
4) Wash 2 x 5 min in ample 2 x SSC, 0.1% SDS at 25°C followed by 2x15 min in 0.5 x SSC, 0.1 SDS at 65°C

4. Immunological Detection

1) After hybridisation and stringency washes, incubate the membrane in 100 mL blocking solution (1% milk powder containing 0.1 M Maleic acid, 0.15 M NaCl, pH 7.5) for 30 min

2) Incubate for 30 min in 20 mL antibody solution (containing anti-Digoxigenin-AP 1:5000 (150mU/mL) in blocking solution)

3) Wash 2x15 min in 100 mL washing buffer (containing 0.1 M Maleic acid, 0.15M NaCl, pH 7.5; 0.3% Tween 20) for 30 min.

4) Equilibrate 2-5 min in 20 mL detection buffer (containing 0.1 M Tris-HCl, 0.1 M NaCl, pH 9.5)

5) Incubate membrane in 10 mL freshly prepared colour substrate solution (containing 40 µL of NBT/BCIP to 2 mL of detection buffer) in an appropriate container in the dark. Do not shake during colour development. The colour precipitate starts to form within a few min and the reaction in usually complete after 16 h. The membrane can be exposed to light for short time periods to monitor colour development

6) Stop the reaction, when desired spot or band intensities are achieved, by washing the membrane for 5 min with 50 mL of sterile double distilled water or with 1xTE buffer

Results can be documented by photocopying the wet filter or by photography.

APPENDIX D

DIGESTION OF DNA WITH RESTRICTION ENDONUCLEASES AND ELECTROPHORESIS OF THE RESTRICTION FRAGMENTS

1. Restriction Enzyme (RE) Digestion of DNA

Enzyme digestions are usually carried out in a 15 μL reaction volume. The amount of enzyme and buffer that must be added to the DNA must be carefully calculated so that at least 1 U of enzyme digests each μg of DNA, and there must be enough DNA to be visualized on an agarose gel.

A rough estimate is that approximately 25 ng of ethidium bromide-stained DNA is visible with the naked eye under the UV light. Typically, 2-3 times as many units are required to ensure complete digestion.

Excessive restriction enzyme in the sample may negatively affect the results; contaminating endonucleases can cleave DNA at other sites, or contaminating exonucleases can damage the ends of restriction fragments. In general, the reactions are carried out as follows.

1. The enzymes are supplied at a concentration of 10 to 12 U/μL

   The **enzymes must be kept on ice at all times** and only sterile pipette tips may be used to pipette the enzymes. The restriction buffers are supplied as 10 × concentrations, but are used at 1 × concentration

2. Pipette the following into a sterile Eppendorf tube

   - 0.1-2 μg DNA
   - 1.5 μL of the appropriate 10 × Restriction Buffer
   - 0.5 μL Restriction enzyme (0.2 - 5 U)
   - Sterile water to a final volume of 15 μL

3. Mix the contents and centrifuge in a microfuge for a few seconds to collect the contents at the bottom of the tube

4. Incubate the reaction mixtures for 1 h at 37°C

5. Add 5 μL of the loading buffer containing bromophenol blue to the reaction mixtures and mix thoroughly. Inactivate the restriction endonucleases by placing the tubes in a water bath at 65°C for 10 min

6. Analyse the reaction mixtures on a 1% agarose gel. The DNA fragments resolved on the gel can be recovered by e.g., glassmilk purification for further use

7. For use other than loading on a gel, samples can be purified by phenol/chloroform extraction and ethanol precipitation. If required, a small aliquot can then be examined electrophoretically before digestion with a second restriction enzyme

Restriction endonuclease cleavage is accomplished by incubating the enzymes with the rDNA under appropriate reaction conditions.

1.1. Materials

- Template rDNA
- 10 x reaction buffer
- Restriction endonucleases (*Hpa*I, *Cfo*I)
- Waterbath or heating block (37°C)

1.2. Method

Pipette the following into a clean 0.5 mL Eppendorf tube

- 15 µL template rDNA
- 2 µL 10 x reaction buffer
- 0.5 µL enzyme
- 2.5 µL sterile water

1. A 20 µL reaction is convenient for analysis by electrophoresis in agarose gels. The amount of DNA to be cleaved can be increased or decreased provided the components are proportionally adjusted
2. Incubate the reaction mixture for 1 h at the recommended temperature (in general, 37°C)
3. Principally, 1U restriction endonuclease completely digests 1 µg of purified DNA in 60 min using the recommended assay conditions. However, crude DNA preparations often require more enzyme and/or more time to complete the reaction
4. Stop the reaction and prepare it for agarose gel electrophoresis by adding 5 µL gel loading buffer
5. Samples are subjected to electrophoresis through a 2.5 % agarose gel for 15 h at a rate of 2 V/cm. ***(See* basic protocol-Appendix B**). Be sure to include a molecular marker (e.g., commercial DNA molecular marker).
6. Agarose gels are stained, visualized on a UV transilluminator and photographed
7. The sizes of the restriction fragments are calculated by comparison to the molecular marker

APPENDIX E

DNA SEQUENCE ANALYSIS

1. Template DNA preparation

The quality and purity of the DNA to be sequenced is very important in ensuring a good sequence reaction. There are many commercial kits on the market that may be used to purify PCR amplicons prior to sequencing. Most of these use a method in which the DNA is bound to a matrix in a short column where it may be washed extensively to remove impurities and then eluted in a suitable volume of water or TE butter. The detailed protocols for use with the different kits are included when purchasing the materials. Although these kits make DNA purification trouble free and are convenient, these add to expense and other options can be taken. We will here describe a very simple technique that does not require specialized reagents and may be used for the purification of PCR amplicons for sequencing analysis. Essentially, the amplicon is electrophoretically separated from any other components in the PCR mix; thereafter the DNA is eluted from the agarose and precipitated with alcohol.

1. Run the PCR and evaluate the quality and integrity of the amplicon
2. If satisfied with the product, run a preparative gel on which you load as much of the amplicon as possible (e.g., four wells filled with 20 µL amplicon each)
3. After electrophoresis and staining, cut the amplicon bands from the gel (using a UV lamp and a scalpel) and transfer to an Eppendorf tube
4. Remove the cap and puncture the bottom of a 500 µL Eppendorf tube with a thin needle. Line the bottom of the tube with a small amount of glasswool, covering the hole
5. Add the DNA containing agarose slice(s) to the tube and place the 500 µL tube inside a 1.5 mL Eppendorf tube
6. Spin the tube at high speed in a microfuge for 2 min, forcing all the fluid in the agarose through the glasswool into the bottom tube. The agarose will be blocked by the glasswool covering the hole in the tube
7. Rinse the agarose twice by adding 200-500 µL of water to the top tube and spinning for 2 min. Most of the DNA will now be in the 1.5 mL tube and may be precipitated with 2 volumes of 100% ethanol, in the presence of 0.3 M sodium Acetate

2. Sequencing reactions (Abi Prism™ Big Dye™ Terminator Cycle Sequencing ready reaction kit)

2.1. Automated sequencing Cycle sequencing reactions

The sequencing reaction can be most conveniently carried out in a thermal cycler, using the principles of PCR, but allowing priming from a single site only:

Assemble the following reaction in a 100 µL tube on ice:

1) 2 µL reaction mix (BigDye™ Terminator Ready reaction mix, Perkin Elmer)
2) 3 µL 5x sequencing buffer
3) 3.2pmol primer
4) 1-14 µL (200–500 ng) template DNA
5) UHQ/sterile dd H_20 to make up to 20 µL

2.2. Cycle sequencing conditions

Place tubes in a thermal cycler, (e.g., GeneAmp PCR System 2000, Perkin Elmer) and set the volume to 20 µl. Use the following programme:

1. Rapid thermal ramp to 96°C
2. Repeat for 25 cycles: 96°C, 10"; 50°C, 5"; 60°C, 4'
3. Rapid thermal ramp to 4°C and hold until ready to purify

2.3. Purification of extension products

1. Transfer each reaction to a 500 µL tube containing 50 µL of 96% ethanol and 2 µl of 3 M sodium acetate, allow to stand for 10 min, centrifuge at 13000 rpm for 30 min
2. Carefully remove all the ethanol (the pellet is invisible)
3. Add 250 µL of 70% ethanol and centrifuge again at 13000 rpm for 30'
4. Remove all the ethanol and dry in a vacuum. This can be stored at -20°C for up to a week

2.4. Using the sequencer

These reactions are run on the ABI377 automated sequencer, or equivalent: The pellets are resuspended in 4 µL of 1:5 blue dextran/ EDTA: deionized formamide, denatured at 95°C and put onto ice. 0.5-1 µL is loaded, depending on the size of the gel. Long Ranger™ gel solution matrix, a ready-to-mix gel solution is used.

---

3. Manual DNA Sequencing Protocol

3.1. Denaturing of double stranded plasmid template

1. Assemble the following in a 1.5mL tube

   a. 500 ng DNA template
   b. dH_20 to 10 µL
   c. 0.1 volume of 2 M NaOH, 2 mM EDTA

2. Incubate for 30 min at 37°C
3. Add 0.1 volume 3 M sodium acetate to stop the reaction
4. Precipitate DNA by adding 2.5 volumes of 100% ethanol and incubating at -70°C for 15 min; centrifuge for 10 min at 12000 rpm
5. Discard the ethanol and wash the pellet with 2.5 volumes of 70% ethanol, centrifuge again for 10 min at 12 000 rpm
6. Discard the ethanol, air-dry the pellet, and re-suspend in 7 µL dH_20

3.2. Annealing of primers to the template DNA

1. Assemble the following reaction in a tube:

   - 7 µL denatured DNA
   - 2 µL 5x sequenase reaction buffer (T7 Sequenase DNA sequencing kit, USB™)
   - 1 µL primer (1.5pmol)
   - dH_20 to 10 µL

2. Incubate the reaction at 65°C, 10 min; 37°C, 10 min; room temperature; then briefly centrifuge the tube contents to the bottom and place on ice until use

3. Prepare 4 separate tubes with different termination mixes (2.5 µL each of ddATP, ddCTP, ddGTP and ddTTP) and mark clearly

3.3. Radiolabelling of DNA

1. Assemble the following in a tube
   - 1 µL (0.1M) DTT
   - 2 µL diluted (1:5) labeling mix.
   - 0.5 µL [α-^{35}S] dATP.
   - 2 µL polymerase (0.3 µL pyrophosphatase + 0.3 µL polymerase + 4.6 µL polymerase dilution buffer).
2. Add the 5.5 µL labeling mix to the 10 µL template mix and incubate for 3 min at room temperature
3. To terminate the reaction, add 3.5 µL to each of the respective termination mix's, incubate at 37°C for 5 min
4. Add 4 µL of stop reagent to each termination tube. This can be kept at -20°C, overnight until gel loading

3.4. Gel assembly and electrophoresis

Prepare a 6 % acrylamide gel as follows

1. Dissolve 42 g of urea, 15mL of 40% acrylamide, and 10mL of 10X TBE in 25mL of dH_20.
2. Adjust the volume to 100 mL with dH_20.
3. Add 100 mL of TEMED and 200 mL of freshly prepared 10% APS
4. Inject the gel between thoroughly cleaned and assembled glass plates and allow to polymerise overnight
5. Assemble the (BIO RAD Sequi-Gen® GT Sequencing Cell) apparatus as per the instruction manual and fill the buffer tank filled with 0.5 X TBE
6. Load stop solution in some of the wells and do a pre-run at 2000V for 1h 40'
7. Load 3 µL of each sample into the wells in the order A-C-T-G, run at 2000V for 2h
8. Load another 3 µL volume and allow to run for a further 2 h
9. Remove the gel and dry on a slab drier (Savant SGD5040) for 1 h, and place in an autoradiography cassette in a darkroom. Leave overnight for exposure
10. Take the autoradiograph from the cassette and dip into the developer for 5', then wash with dH_2O. Dip in fixer for 5' and wash again with dH_2O, hang out to dry

4. References

1. Sanger, F, Niklen, S, and Coulson, A. R (1977) *Proc. Nat. Acad. Sc.. USA* **74**, 5463-5467
2. Sambrook, J, Fritsch, E F, and Maniatis, T (Eds) (1989) *Molecular Cloning- A laboratory manual*. 2nd Edition, CSH Laboratory press

APPENDIX F

AN EXAMPLE OF A RT-PCR PROTOCOL

Aim

To prepare a complementary DNA (cDNA) copy of the rabies virus pseudogene from viral RNA and to amplify this cDNA by PCR.

1.1. Materials

1. Rabies virus RNA or virus-infected tissue

2. Oligonucleotide primers:

G (+) [77 OD/mL]: Anneals to the (-) strand of the viral genome at nucleotides 4665-4687
L (-) [59 OD/mL]: Anneals to the (+) strand of the viral genome at nucleotides 5543-5566

3. Reverse Transcriptase: M-MLV Reverse Transcriptase

4. Ribonuclease inhibitor: 0.5 U/μL reaction

5. TRIReagent (Phenol-base): Molecular Research Centre, USA.

1.2. Methods

1.2.1. RNA extraction

Rabies virus RNA is most typically obtained by extraction from virus-infected mouse brain material as follows:

1. Transfer 500 μL virus-infected tissue to an Eppendorf tube
2. Add 1 mL of TRIReagent, mix and incubate at RT for 5 min
3. Add 200 μL chloroform, shake vigorously, incubate at RT for 15 min
4. Spin at high speed (13000 g) in microfuge for 10 min
5. Total RNA is precipitated from the supernatant by addition of 500 uL of isopropanol
6. RNA is pelleted by high-speed spin (15 min.) and the pellet washed with 70% ethanol
7. The RNA is dissolved in 50 μL of RNA'se free H_2O, and the concentration determined by spectroscopy

1.2.2. cDNA Synthesis

1. Combine in an Eppendorf tube

4 μL RNA (0.5-1 μg)
1 μL G (+) [100ng/mL]

Hybridise at 65°C for 3 min. Place immediately on ice.

2. Add the following

- 1 μL Ribonuclease inhibitor
- 2 μL 5x buffer
- 1 μL 10 mM dNTP mix
- 1 μL DTT
- 2 μL Reverse transcriptase enzyme

3. Incubate at 42°C for 90 min.
4. Dilute to 100 μL with ddH_20

1.2.3. PCR

1. Combine the following in a 0.5 mL Eppendorf tube and adjust the volume to 100 µL with UHQ H_20:

- 10 µL cDNA
- 10 µL 10x Buffer (500mM KCl; 100 mM Tris.HCl pH9; 1% Triton X-100)
- 6 µL 25 mM $MgCl_2$ (1.5 mM final)
- 10 µL 1 mM dNTP mix
- 100 pmol G(+) (8 µL of a 100 ng/µL stock)
- 100 pmol L(-) (8 µL of a 100 ng/µL stock)

2. Overlay the reaction mixture with mineral oil and denature at 95°C for 3 min.
Add 0.25U Taq polymerase, ensuring that the pipette tip is submerged well below the oil layer.

3. Conduct the PCR with the following cycles:

- 1 cycle 94°C; 1 min
- Denaturation: 94°C, 1 min
- Annealing: 42°C, 1.5 min
- Chain elongation: 70°C, 2 min (30 cycles)

4. Analyse the PCR products by agarose gel electrophoresis **(Appendix B)**

APPENDIX G (i)

CALIBRATION AND MAINTENANCE OF MECHANICAL PIPETTES

This SOP describes the calibration and maintenance of mechanical pipettes used in the PCR laboratory. Special reference to Eppendorf micro-pipettes is made.

1. Safety/precaution

Air drafts and temperature fluctuations should be eliminated prior to pipette calibration.

2. Equipment

Model:	Eppendorf
Manufacturer:	Eppendorf
Date of purchase:	2001
Initial condition:	New
Location:	PCR Laboratory 1, 2 and 4

3. Materials and accessories

- Analytical balance (micrograms)
- HPLC Water
- Weigh boats

3.1. Gravimetric Method

- The calibration procedure described below, for using 20 replicate measurements, should be performed quarterly and on new pipettes before use

- For monthly checks, use the same procedure except that only 2 readings are necessary at the appropriate volumes

- All specifications apply to measurements of water at an ambient temperature range of 18 to 25°C

- Due to evaporation of water, it is imperative that there be no delays in taking the readings, especially when calibrating volumes 100 μL

- Water evaporation can lead to poor coefficient of variation values due to weight-over-time-changes

- Although speed is imperative, accuracy should not be sacrificed
-

  1. Obtain a beaker of distilled, deionised water
  2. Place a small beaker or weighing dish on a precision microgram analytical balance
  3. Adjust the weight to zero with the tare knob
  4. Put a pipette tip on the pipette to be calibrated. Rinse the tip 2 to 3 times with water, withdraw a sample, then dispense the water into the empty container on the balance pan
  5. Record the weight **(see Appendix A),** reset the pan arrest, and deliver a second aliquot of water into the container on the balance pan.
  6. Record the new weight
  7. Repeat this process 20 times recording the weight each time

3.1.1. Calculations

To determine the weight of the water in the container after each pipetting, subtract the weight of the water in the container before delivery from the pipette from the weight of the water in the container after delivery from the pipette.

The difference will represent the weight of the water delivered by the pipette.

To calculate the mean pipetting volume, add up all the numbers in column 3 (Appendix A) and divide by 20.

1. Calculate the standard deviation (SD) of the values in column 3.
2. Calculate precision using the coefficient of variation (relative %):

$$= \frac{SD \times 100}{Mean}$$

3. Calculate accuracy (Mean Error):

Mean Volume - Volume displayed on pipette

4. Calculate % Mean Error:

$$\frac{\text{Mean Volume - Volume Displayed}}{\text{Volume Displayed}} \quad x\ 100$$

3.2. Guidelines for accuracy and precision measurements of pipettes

Volume (µL)	Accuracy (Relative %)	Precision (Relative %)
0.5	≤5.0	≤3.0
1	≤2.5	≤1.5
5	≤1.5	≤0.6
10-20	≤1.0	≤0.5
50-100	≤1.0	≤0.3
200-1000	≤0.8	≤0.2

4. Cleaning and maintenance

- Digital pipettes do not require maintenance under normal use
- For the newer models, all parts of the pipette can be cleaned with a soap solution or isopropanol
- After cleaning, rinse with distilled water and dry
- The piston should then be lubricated lightly with silicone grease
- See the trouble-shooting section of the manual if any of the following problems are encountered: droplets on the inside of the tip, dripping pipette or erratic movement of the control button
- If replacement parts are needed, refer to the section of the manual detailing the replacement of these parts

For the older models, if the nose cone gets clogged, it can be cleaned with distilled water as described in the older Eppendorf Instruction Manual.

If the piston needs cleaning, it can be wiped with a damp cloth as also described in the Eppendorf manual.

If cleaning the nose cone and/or piston does not solve pipetting problems, the pipette will probably have to be retired since Eppendorf no longer repairs or carries replacement parts for the older models.

5. Document all maintenance in the maintenance section of the pipette calibration logbook.

Sample calibration log sheets-initial and quarterly pipette calibration

Pipette Brand_______________ Pipette Number_______ Pipette volume_______

Date______________________ Technician______________________________

Mean Pipetting Volume =	
Accuracy (Mean Error) =	
Coefficient of Variation =	
Standard Deviation =	
% Mean Error =	

APPENDIX G (ii)

STANDARD OPERATING PROCEDURE FOR USING MICROPIPETTES

This SOP describes the procedure for using micro pipettes

1. Safety/precaution

Always wear gloves and protective clothes when holding Eppendorf pipettes to avoid handling contamination. Never lay the pipette down with liquid in the tip, as this can flow into the pipette.

2. Equipment

Model.	0.1-2.5 μL	dark gray
	0.5-10 μL	light gray
	10-100 μL	yellow
	100-1000 μL	blue

Manufacturer: Eppendorf

Date of purchase: 2001

Initial condition: New

Location: PCR Laboratory 1, 2 & 4

3. Materials and Accessories

Pipette tips

4. Operational procedure

Each pipette is equipped with a comfortable handgrip, a disposal button, a delivery button, and a grooved wheel for volume delivery adjustment.
The delivery button, when pushed, is equipped with two "STOP" points, noticeable by the feel of gentle resistance.

The first STOP is to obtain the proper delivery amount

The second to insure complete delivery

The volume to be delivered is adjusted with the grooved wheel, and read with reference to the numbers and vernier scale on the side of the grip.

1. Select desired volume
2. Select appropriate tip

4.1. Filling

For optimal precision and accuracy

- Volumes > 10 mL-pre-rinse 2-3 times for each new tip (this involves aspirating and dispensing liquid several times)
- Volumes < 10 mL - DO NOT pre-rinse (rinse tip AFTER dispensing to ensure whole volume dispensed)
- Make sure tip is securely attached
- Hold pipette upright
- When aspirating, try to keep the tip at a constant depth below the surface of the liquid.
- Glide control button *slowly* and *smoothly* (electronic pipettes perform this step for you).
- When pipetting viscous liquids (i.e., serum, blood), leave tip in the liquid for 1-2 seconds after aspirating before withdrawing it
- Once there is liquid in the tip, NEVER lay pipettes on their sides

4.2. Dispensing

Hold tip at an angle, against the inside wall of the vessel/tube if possible.
Glide control button slowly and smoothly (electronic pipettes perform this step for you).

APPENDIX H

STANDARD OPERATING PROCEDURE FOR USING CLOCKS AND TIMERS

This SOP describes the procedure for using clock/timers

1. Safety/precaution

Always hold it with care to prevent it from falling.

2. Equipment

Model: CE

Manufacturer:
Date of purchase
Initial condition: New
Location PCR Laboratory 2 & 4

3. Method

3.1. Battery installation

To insert button cell battery (Maxell LR44, Union Carbide A-76 or equivalent), remove battery door on the back cover of case and insert battery in direction indicated.

3.2. To set clock

1. Adjust the sliding switch to the clock position, 12:0000 will appear upon power-up.
2. Press "ENTER" for more than 2 s to set clock time. The clock display will be flashing
3. Press "HR", "MIN", "SEC" buttons to set actual time. To fast advance, press and hold the desired button
4. Press "CLEAR" to select 12/24 h format
5. Press enter to exit time setting. The unit will automatically return to normal time display if no input is detected for 3 s

3.3. To start count down

1. Adjust sliding switch to "COUNT DOWN" position, display shows "HOUR", "MINUTE", SECOND" together with "COUNT DOWN" indicator
2. Press "HR", "MIN", "SEC" button to set desired time cycle

count down
18: 3808

3. Press "START/STOP" to start count down, you can temporary stop or re-start timing cycle by pressing the "START/STOP" button

3.4. To start count up

1. Adjust the sliding switch to "COUNT UP" position
2. Press the "START/STOP" button to start timer from 0:0000. Alternatively, the timer can also count from any desired time set by "HR", "MIN" and "SEC" keys

count up
0:00 00

3. The timer can be temporarily stopped by pressing "the "CLEAR" button

4. Timer can count up to 99 hr 59 min 59 s and stop at 0: 0000. A 60-s alarm signal and "TIME'S UP" flag will sound and flash simultaneously

5. Press any button to stop alarm

3.5. Program timer

The unit allows 4 different count down timer settings to be stored in memory (T1, T2, T3 and T4) with respective alarm beeping sequence as below.

count down	"T1" alarm sound: "BEEP in 1 second
T1	"T2" alarm sound: "BEEP BEEP" in 1 second
0: 00 01	"T3" alarm sound: "BEEP BEEP" BEEP in 1 second
	"T4" alarm sound: "BEEP BEEP "BEEP "BEEP in 1 second

APPENDIX I

THEORETICAL BACKGOUND TO PH, BUFFERS AND CONCENTRATION OF SOLUTIONS

1. pH:

$$pH = -\log [H^+]$$

$$2H_2O \underset{k_2}{\overset{k_1}{\rightleftharpoons}} H_3O^+ + OH^-$$

Net reaction

$$H_2O\ (l) \underset{k_2}{\overset{k_1}{\rightleftharpoons}} H^+\ (aq) + OH^-\ (aq)$$

$$K_{eq} = k_1 / k_2$$

$$K_{eq} = [H^+][OH^-] / [H_2O]$$

Water dissociates to a very small extent, therefore the concentration of undissociated water remains constant.
The equilibrium constant of water (K_w) is determined by $[H^+]$ and $[OH^-]$

$$K_w = [H^+][OH^-] = 1 \times 10^{-14}$$

Pure H_2O has equal amounts of H^+ and OH^- ions

$$[H^+] = 10^{-7}$$

More convenient to use the pH scale

$$pH = \log 1/[H^+] = -\log [H^+] = -\log 10^{-7} = 7$$

2. Buffers

Definitions:
In aqueous solutions, acids and bases react with water to give H_3O^+ and OH^-

- Acid: produces H_3O^+ on reaction with water
- Base: produces OH^- on reaction with water
- A solution is buffered if the pH resists change upon addition of a strong acid (H_3O^+) or base (OH^-)
- A buffer contains equal concentrations of a weak acid/base and it's conjugated base/acid

Two species are therefore required in a buffer solution:

- Acid (reacts with added hydroxyl ions)
- Base (reacts with added hydronium ions)

2.1. pK_a values for buffer solutions

The values of pK_a for buffers are determined by titration.

The pK_a of a weak acid or base generally indicates the pH of the center of the buffering system. Always work within one pH unit below or above the pK_a value of a given buffer.

Weak acid	Conjugate base	Acid K_a	pK_a	Useful pH range
Phthalic acid	Hydrogen phthalate	1.3×10^{-3}	2.9	1.9 – 3.9
Acetic acid	Acetate	1.8×10^{-5}	4.7	3.7-5.8
Dihydrogen phosphate	Hydrogen phosphate	6.2×10^{-8}	7.2	6.2 – 8.2
Hydrogen phosphate	Phosphate	3.6×10^{-13}	12.4	11.3 – 13.3

For example: Acetic acid and acetate ion

$$CH_3COOH\ (aq) + OH^-\ (aq) \rightleftharpoons CH_3COO^-\ (aq) + H_2O\ (l)$$

$$HAc \underset{k2}{\overset{k1}{\rightleftharpoons}} H^+ + Ac^-$$

$k_1\ [HAc] = k_2\ [H^+][Ac^-]$

$k_1 / k_2 = [H^+]\ [Ac^-] / [HAc]$

$k_1 / k_2 = K_a$ **Equilibrium constant**

$[H^+] = K_a\ [HAc] / [Ac^-]$

$-\log [H^+] = -\log K_a - \log\ [HAc] / [Ac^-]$

$pH = -\log [H^+]$ and $pK_a = -\log K_a$

$pH = pK_a - \log [HAc] / [Ac^-]$

OR

$pH = pK_a + \log [Ac^-] / [HAc]$ **(Henderson-Hasselbach equation)**

2.2. Buffer capacity and range

The buffer capacity of a solution is the number of equivalents of a strong acid/base that causes 1L of the buffer to undergo a 1 unit change in pH.
Effect of addition of a strong acid to a non-buffered system:
100 mL of 1×10^{-2} M HCl are added to 1 L of 1 M NaCl at pH 7, the $[H^+]$ of the resulting 1.1 L of solution can be calculated as follows:

$[H^+] * Vol = [H^+]_0 * Vol_0$
$[H^+] * 1.1 = 1 \times 10^{-2} * 0.1$
$[H^+] = 9.09 \times 10^{-4}$
$pH = 3.04$

A pH change of approximately 4 pH units resulted

Now we substitute the NaCl solution for a hypothetical buffer at pH 7, with a concentration of 1 M and a pK_a of 7.

$$[HA] = [A] = 0.5\ M$$
$$pH = pK_a + \log [Ac^-] / [HAc]$$
$$pH = 7 + \log 0.5 / 0.5$$
$$pH = 7$$

When 100 mL of 1×10^{-2} M HCl is added to this system, 1×10^{-3} moles of A^- are converted to 1×10^{-3} moles of HA

$$pH = 7.000 + \log (0.499/1100) / (0.501/1100)$$
$$pH = 7.000 - 0.002$$
$$pH = 6.998$$

The buffer system successfully resisted a change in pH

Effective buffering is obtained within the range from

$[A^-]/[HA] = 0.1 - 10$

The maximum buffer capacity is when the concentration conjugated acid/base is equivalent ($[A^-]=[HA]$).

Important

- An acid at low pH has a large pK_a due to complete dissociation.
- Buffer at low concentration - complete dissociation therefore, low buffer capacity

2.3. Selection of a buffer for biological experiments

pK_a between 6 and 8. High solubility in aqueous systems

Exclusion by biological membranes
Minimal salt effects
Minimal effects on dissociation due to concentration, temperature and ionic composition
Well defined, or non-existent interactions with mineral cations
Chemical stability
Insignificant light absorption between 240 and 700 nm
Easily available in pure form
The pK_a of the buffer should lie within 1 unit of the desired pH
The flat region of the titration curve is where HA:A^- change very slowly and where resistance to change in pH is optimal

2.4. Zwitter ionic buffers

At *physiological pH* the alpha-carboxyl and the alpha-amino groups are ionised with the deprotonated carboxyl group bearing a negative charge and the protonated amino group a positive charge.

- An amino acid in its dipolar state is called a zwitter ion.
- Isoelectric point is the pH where there is no net migration of the zwitter ion in an electric field

2.5. Concentrations of solutions

Liquids and solids dissolved in liquids

Solution concentrations can be divided into two types based on

- volume
- weight

2.6. Volume concentrations

2.6.1. Molar
A solution is one molar, 1 M, when it contains one mole of solute in a liter of solution.

2.6.2. Normal
A solution is one normal, 1 N, when it contains one gram-equivalent weight of solute in a liter of solution.

2.6.3. Weight / volume percent (w/v)

A solution is 1 % (w/v) when it contains 1 g of solute per 100 mL of solution

2.6.4. Molarity (M) = moles of solute / liters of solution

2.7. Moles

The amount of substance that contains as many particles as there are atoms in 0.012 kg of C^{12} isotope

6.022×10^{23} Avogadro's constant

A mole always contains the same number of particles, no matter what the substance

---

2.8. Molar mass: M (g/mol)

Numerically equal to the atomic weight in atomic mass units
Mass of 1 mole substance

On the periodic table of elements

16[1]	[3]32.07
	S[2]

1) Atomic number (Z)
2) Element symbol
3) Relative atomic mass (A)

A = protons
Z = protons + neutrons

Moles	→	Grams	(Grams = moles x molar mass)
Grams	→	Moles	(moles = grams / molar mass)

i.e. How many grams in 2.5 moles of S
2.50 x 32.07 g/mol
= 80.175 g S

How many moles in 454 g of Si
454 g / 28.09 g/mol
= 16.16 mol Si

Examples:

$NaHCO_3$	26.3 g
Dissolved in water	200 mL

2.8.1. Calculate the molarity

n = c/V
n = mol
c = concentration (mol/litre)
V = volume

$M (NaHCO_3) = 22.99 + 1.01 + 12.01 + 3 \times 16$
$= 84$ g/mol

g = c x Mr x V
c = g / (Mr x V)
c = 26.3 / (84 g/mol x 0.2 L)
c = 1.57 M

2.9. Solutions of known concentration

e.g., Prepare 2 L of a 1.5 M Na_2CO_3 solution
Moles required ?

2.0 L * (1.5 moles /L) = 3 moles Na_2CO_3
moles to gram: 3 x (106 g / mol)
$= 3.2 \times 10^2$ g Na_2CO_3
= 320 g

2.10. Dilutions

$C_1V_1 = C_2V_2$

C_1 = start [ ]
V_1 = start volume
C_2 = final [ ]

V_2 = FINAL VOLUME

- Dilution of primers upon arrival
- Storage
- Calculation of MM in Dalton
  MM of deoxynucleotide = 330 Da
  MM of DNA basepair = 660 Da
  MM of ds DNA = bp x MM

2.11. Conversion of µg to pmol

pmol of ds DNA
$= \mu g \times (10^6\ pg\ /\ 1\ \mu g) \times (1\ pmol/660\ pg) \times 1/\ Nbp$
$= \mu g \times 1.515\ /\ Nbp$

pmol of ss DNA
$= \mu g \times (10^6\ pg\ /\ 1\ \mu g) \times (1\ pmol\ /\ 330\ pg) \times 1/Nb$
$= \mu g \times 3.03\ /\ Nb$

2.12. Conversion of pmol to µg:

ds: $\mu g = pmol \times Nbp \times 6.6 \times 10^{-4}$

ss: $\mu g = pmol \times Nb \times 3.3 \times 10^{-4}$

1 A_{260} unit ds DNA = 50 µg / mLH_2O`

1 A_{260} unit ss DNA = 33 µg/mLH_2O

APPENDIX J

GENERAL BUFFER AND STOCK SOLUTIONS

1. Bromophenol blue, 10%

Mix 5 g of bromophenol blue in 50 mL of H_2O. Store at room temperature.

2. EDTA (ethylenediamine tetraacetic acid, disodium salt), 0.5 M

1. To 300 mL of H_2O in beaker add 93.05 g EDTA-$Na_2.2H_2O$
2. Mix well. Add 10 N NaOH to pH 8.0
3. (Will not go into solution until about pH 7.)
4. Add H_2O to 500 mL

3. Ethidium bromide solution, 10 mg/mL

Used to visualize DNA; emits fluorescence when excited at 250-310 nm.
M.W. 394.32.

1. Add 0.2 g of ethidium bromide to 20 mL of H_2O
2. Mix well. Store at 4°C in dark (brown or foil covered bottle)
3. Handle with gloves and avoid inhalation

4. Formamide, deionized.

1. Add 50 mL of formamide to 5 g of a mixed-bed resin (e.g., Bio-Rad AG 501-X8)
2. Stir gently at 4°C and filter twice through Whatman No. 1 filter paper
3. Aliquot and store at –20°C

5. Gel loading buffer.

Used to load DNA samples on gels

To make	10 mL	Final conc.
Glycerol	5 mL	50%
40X TAE buffer	250 µL	1X
Bromophenol blue, saturated	1 mL	1%
Xylene cyanol, 10% suspension	1 mL	1%
H_2O	2.75 mL	

Typically, make 10 mL, divide into 1-mL aliquots, and store at –20°C.

---

6. IPTG (isopropyl-ß-D-thiogalactopyranoside) 100 mM in H_2O

23.8 mg IPTG in 1 mL H_2O

---

7. $MgCl_2$, 1 M

1. Add 20.3 g of $MgCl_2 \cdot 6H_2O$ to a final volume of 100 mL of H_2O
2. Autoclave to sterilize

---

8. Sodium acetate, 3 M

40.8 of sodium acetate · $3H_2O$ in a final volume of 100 mL of H_2O
pH adjusted with acetic acid, as indicated; typically 7.0 for use with DNA, and 6.0 and sterile filtered for RNA

---

9. SDS (sodium dodecyl sulfate), 10%

100-g bottle brought to 1 L with H_2O
Use whole bottle to avoid breathing dust

10. SS-phenol (salt-saturated phenol)

1. 1 lb bottle phenol (e.g., Mallincrodt)
2. 100 mL 2 M Tris, pH 7.4
3. 130 mL H_2O
4. Heat at 37°C carefully until phenol is dissolved.
5. Remove upper aqueous phase.
6. Add: 100 mL 2 M Tris, pH 7.4
7. 25 mL m-cresol (e.g., Kodak)
8. 1 mL ß-mercaptoethanol
9. 500 mg 8-hydroxyquinoline

a. Store aqueous and phenol layers together in hood at room temperature
b. Use the yellow phenol solution for extraction
c. Note that the phenol will remain as the top layer in the stock bottle for easy removal
d. When used to extract DNA, the SS-phenol will be typically the bottom layer

11. TAE buffer, 40 X

To make 1 liter

a.	Tris base (1.6 M)	193.6 g
b.	Na acetate.3H_2O (0.8 M)	108.9 g
c.	EDTA-Na_2.2H_2O (40 mM)	15.2 g
d.	pH to 7.2 with acetic acid	
e.	H_2O to make	1 liter

12. TBE buffer, 20 X

To make 1 liter

a.	Tris base (1 M)	121 g
b.	Boric Acid (1 M)	61.7 g
c.	EDTA-Na_2.2H_2O (20 mM)	7.2 g
d.	H_2O to make	1 liter

Store at room temperature. Remake if major residues appear in solution.

13. TE buffer (Tris/EDTA)

To make 100 mL:

- Tris, 10 mM, pH 7.4 — 0.5 mL of 2 M stock
- 0.1 mM EDTA, pH 8 — 20 µL of 0.5 M stock
- H_2O to make — 99.3 mL

Store at room temperature

APPENDIX K

GENERAL REAGENTS AND SUPPLIERS

1. Items list

Item	Supplier
Electrophoresis	
Agarose	MERCK/Sigma/Strategene
100bp DNA Ladder	Promega/Sigma/Sterilab/Strategene
Acetic acid	MERCK/Sigma
Agarose LE	Roche/Promega/Seakem/Strategene
Boric acid	MERCK/Sigma
Ethylenedinitnlo)tetraacetic acid	AEC-Amersham
Ethidium bromide	Sigma/MERCK
Metaphor Agarose	Sterilab/AEC-Amersham
Buffers	
Tris	MERCK/Sigma/Roche/AEC-Amersham
Tris HCl	MERCK/Sigma/Roche/AEC-Amersham
PCR Reagents	
dUTP, PCR Grade	Roche/PE Biosystems/ Promega/Stratagene
100 mM dNTP set	Roche
High Quality H2O (nuclease free)	Promega/MERCK/AEC-Amersham
Human Placenta Ribonuclease Inhibitor	AEC-Amersham
Random Hexamers	Promega
Uracil-DNA Glycosylase	Roche
Cloning	
0.2 mL PCR tubes, with flat cap	Qiagen/Kartell/Elkay
1.5 mL centrifuge tubes	Promega
1.8 mL cryotubes	LASEC
1000 µL filter tips	Promega
100 µL filter tips	Promega
10mL serological pipettes	NUNC
10 µL filter tips	Promega
1 mL disp. Syringe with 29g needle	Omnimed
250 mL Oakridge tubes	LASEC

2 mL centrifuge tubes	AEC-Amersham
2-Propanol	MERCK
50 mL oakridge tubes	LASEC
50 mL, self standing tube	AEC-Amersham
5 mL serological pipettes	NUNC
6 place micro pipette stand	Promega
6 well dishes	LASEC
8 mL tubes with screw caps	AEC-Amersham
AB Gene 0.2mL Micro-tube	Qiagen
Absolute ethanol	MERCK/Sterilab
Biosolve	Sigma/MERCK
Cameo 0.22, 25mm filters	Sigma/Pierce
Cell culture work	
Cryotubes	LASEC
Disp. Cuvettes	Lab. & Scientific equipm
Easy flask 175 cm2	NUNC
Easy flask 25 V/C	NUNC
Easy flask 75 V/C	NUNC
Labware	
Lamda DNA	Promega/Sterilab
Modifying enzymes	Roche/Promega
Nuclon Tube & screw cap	AEC-Amersham
PGEM-T Easy vector system	Promega
Potassium Acetate	Sigma/MERCK
Restriction Enzymes	Promega/Roche/AEC-Amersham/New EnglandBiolabs
Kits	
Sequencing buffer	Applied Biosystems
GenElute TM Gel purification Kit	Sigma
Equipment	
Analytical balance	Heraeus
Centrifuge	LASEC
Electrophoresis apparatus	Bio-Rad
Flow cabinet	Local suppliers
Freezers	Local suppliers
Fridges	Local suppliers
Heating blocks	MERCK/COPE
Micropipettes	MERCK – eppendorf
Microwave	Local suppliers
PCR machine	PE Biosystems
pH meter	Heraeus/Separtations

Vortex	Heraeus
Nucleic acid extraction	
IAQuick Gel extraction kit	Qiagen
QIAamp Viral RNA kit	Qiagen
QIGEN Giga kit (5)	Qiagen
QIquick Gel extraction kit	Qiagen
Tri-pure	Roche
Tri-reagent	Sigma
Tri-zol	Qiagen
Enzymes	
Restriction Enzymes	Roche/AEC-Amersham/Sigma/Promega
FastStart Taq Polymerase	Roche
HotStarTaq DNA Polymerase	Qiagen
Moloney Murine Leukemia Virus Reverse Transcription	Promega
Takara Ex Taq	Pierce
Oligonucleotides	Promega/Roche/Lab. Spec. Service

2. Suppliers

SUPPLIERS LIST			
AEC-Amersham	Cust.servex@eu.amershambiosciences.com	+43 (1) 982 3826	www.amershambiosciences
Bibby Sterilin - Sterilab	Marketing@bibby-sterilin.fr	+331 6445 1313	www.bibby-sterilin.com
Bio-Rad	Sadtler.europe@bio-rad.com	01-809 5555	www.discover.bio-rad.com
A.J. Cope & Sons Ltd	The Oval Hackney Road London, E2 9DU, UK	0171 729 2405	
Elkay	Sales@elkay-uk.co.uk	353 91 572266	www.elkay-uk.co.uk
Heraeus	Info@kendro-lab.fr	+33 (1) 6918 7777	www.heraeus-instruments.d
Invitrogen - Lab. Spec. Services	Eurotech@invitrogen.com	0800 91 83 92	www.invitrogen.com
Kartell	kartell@kartell.it	02 900121	www.kartell.it
MERCK	Service@merck.de	+49 6151 72 0	www.merck.de
NUNC	Infociety@nunc.dk	+45 4631 2000	www.nuncbrand.com
PE Biosystems	France Applied Biosystems 25, Av de la Baltique B.P. 96 91943 Courtalboeuf, Cedex	33 (0) 1 69 59 85 85	www.perkin-elmer.com/ab
Pierce	De.info@pebio.com	0882 991 25 650	www.pebio.com
Promega	catalys_custserv@catalys.promega.com	800 356 9526	www.promega.com
Qiagen	info@westburg.nl	01-60 92 09 20	www.qiagen.com
Roche	Biochem.fr@roche.com	04 7676 3087	http://biochem.roche.com
Sigma	Genosys@genosys.co.uk	+44 (0) 1223 839000	www.sigma-aldrich.com
Strategene	Techservice@stratagene.com	(800) 424 5444x2	www.stratagene.com

APPENDIX L

BIOINFORMATICS FREEWARE, SOFTWARE AND ANALYSIS PROGRAMMES

1. Molecular biology tools online

- **www.ncbi.nlm.nih.gov**

1.1. What does ncbi stand for? National Centre for Biotechnology Information

- A national resource for molecular biology information
- Creates public databases
- Conducts research in computational biology
- Develops software tools for analysing genome data, disseminates biomedical information all for the better understanding of molecular processes affecting human health and disease

---

1.2. Databases and Software

- GenBank NIH genetic sequence database.An annotated collection of all publicly available DNA sequences.17 billion bases from over 100,000 species
- GenBank (at NCBI) + DNA DataBank of Japan (DDBJ) + European Molecular Biology Laboratory (EMBL) comprise the International Nucleotide Sequence Database Collaboration

NCBI supports and distributes a variety of databases for the medical and scientific communities.

These include:

- The Online Mendelian Inheritance in Man (OMIM)
- The Molecular Modeling Database (MMDB) of 3D protein structures
- The Unique Human Gene Sequence Collection (UniGene)
- A Gene Map of the Human Genome
- The Taxonomy Browser and The Cancer Genome Anatomy Project (CGAP), in collaboration with the National Cancer Institute

1.2.1. Entrez

- NCBI's search and retrieval system that provides users with integrated access to sequence, mapping, taxonomy, and structural data. also provides graphical views of sequences and chromosome maps
- A powerful and unique feature of Entrez is the ability to retrieve related sequences, structures, and references

1.2.2. PubMed (journal literature)

A Web search interface that provides access to over 11 million journal citations in MEDLINE and contains links to full-text articles at participating publishers' Web sites

1.2.3. BLAST

- A program for sequence similarity searching developed at NCBI, and is instrumental in identifying genes and genetic features
- Can execute sequence searches against the entire DNA database in less than 15 s

Additional software tools provided by NCBI include:

- Open Reading Frame Finder (ORF Finder)
- Electronic PCR
- The sequence submission tools, Sequin and BankIt

All of NCBI's databases and software tools are available from the WWW or by FTP. NCBI also has email servers that provide an alternative way to access the databases for text searching or sequence similarity searching.

1.3. Searching GenBank for a sequence

- Go to the website: **www.ncbi.nlm.nih.gov**
- At the NCBI toolbar: click on the nucleotide dropdown menu and select nucleotide
- Type in the key word for your search e.g. *Rift valley fever virus.*
- The page thereafter will give you information on all the available nucleotide sequences for *Rift valley fever virus*, as well as an accession number that is unique for that particular sequence
- Select the particular sequence of choice by clicking on the accession number for that sequence
- The results page will contain information on:
    - What sequence it is
    - The organism
    - The authors of the sequence
    - Journal reference if available
    - The nucleotide sequence
    - The amino acid sequence

1.4. Searching GenBank using an accession number

- Go to the website: **www.ncbi.nlm.nih.gov**
- At the NCBI toolbar: click on the nucleotide dropdown menu and select nucleotide.
- Type in the key word for your search e.g. *D10065*
- The page thereafter will give you information on the accession number that is unique for that particular sequence
- Select the accession number by clicking on it
- The results page will contain information on:
    - What sequence it is
    - The organism
    - The authors of the sequence
    - Journal reference if available
    - The nucleotide sequence
    - The amino acid sequence

1.5. What is BLAST?

- Basic Local Alignment Search Tool
- Method for rapid searching of nucleotide and protein databases
- A tool to view sequences aligned with each other or to find homology, but a program to locate regions of sequence similarity with a view to comparing structure and function

1.5.1. Selecting the BLAST program

The BLAST search pages allow you to select from several different programs. Below is a table of these programs:

Program	Description
blastp	Compares an amino acid sequence against protein sequence data base
blastn	Compares a nucleotide query sequence against a nucleotide sequence database
blastx	Compares a nucleotide quert sequence translated in all reading fames against a protein sequence database. You could use this option to find potential translation products of an unknown nucleotide sequence
tblastn	Compares a protein query sequence against a nucleotide sequence database dynamically translated in all reading fames
tbalstx	Compares the six-frame translations of a nucleotide sequence against the six-frame translations of a nucleotide sequence database. Please note that the tblastx program cannot be used with the nr database on the BLAST web page because it is computationally intensive

1.5.2. If your sequence is nucleotide

Length	Database	Purpose	BLAST Program
20bp or longer	**Nucleotide**	Identify the query sequence	MEGABLAST
			Standard BLAST (blastn)
		Find sequence similar to query sequence	Standard BLAST (blastn)
		Finds similar proteins to translated query in a translated database	Translated BLAST (tbalstx)
	Protein	Find similar proteins to translated query in a protein database	Translated BLAST (blastx)
7-20bp	**Nucleotide**	Find primer binding sites or map short contiguous motifs	Search for short nearly exact matches

1.5.3. If your sequence is protein

Length	Database	Purpose	BLAST program
15 residues or longer	Protein	Identify the query sequence or find protein sequences similar to query	Standard Protein BLAST (blastp)
		Find members of a protein family or build a custom position specific core matrix	PSI-BLAST
		Find proteins similar to the query around a given pattern	PHI-BLAST
	Conserver proteins	Find conserved domains in the query	CD-search (RPS-BLAST)
	Conserved domains	Find conserved domains in the query and identify other proteins with similar domain properties	Domain Architecture Retrieval Tool (DART)
	Nucleotide	Finds similar proteins in a translated nucleotide database	Translated BLAST (tblastn)
5-15 residues	Protein	Search for peptide motifs	Search for short, nearly exact matches

1.5.4. BLAST a primer sequence

- Go to the website: **www.ncbi.nlm.nih.gov**
- At the NCBI toolbar: click on *BLAST*
- On the BLAST page, under the title of nucleotide BLAST, select *search for short nearly exact matches*
- On the following page type in the primer sequence in the search box and click on *BLAST*.
- On the formatting BLAST page, click on format

1.5.5. BLAST result

- ♦ **The BLAST report header.** The top line gives information about the type of program, the version, and a version release date. The research paper that describes BLAST is then cited, followed the request ID, the query sequence definition line, and a summary of the database searched
- ♦ **Graphical overview of BLAST results.** The query sequence is represented by the numbered red bar at the top of the figure, Database hits are shown aligned to the query, below. Of the aligned sequences, the most similar are shown closest to the query
- ♦ **One-line descriptions in the BLAST report**

Each line is composed of four fields:

a) The gi number, database designation, Accession number, and locus name for the matched sequence, separated by vertical bars

b) A brief textual description of the sequence, the definition. This usually includes information on the organism from which the sequence was derived, the type of sequence (e.g., mRNA or DNA), and some information about function or phenotype. The definition line is often truncated in the one-line descriptions to keep the display compact.

c) The alignment score in bits. Higher scoring hits are found at the top of the list

d) The E (expect) value, that provides an estimate of statistical significance

1.5.6. A pair-wise sequence alignment from a BLAST report

- The alignment is preceded by the sequence identifier, the full definition line, and the length of the matched sequence, in amino acids
- Next comes the bit score (the raw score is in parentheses) and the E value
- The following line contains information on the number of identical residues in this alignment ("Identities"), the number of conservative substitutions ("Positives") and, if applicable, the number of gaps in the alignment
- Finally the actual alignment is shown, with the query on top and the database match labeled as 'Sbjct', below. The numbers at left and right refer to the position in the amino acid sequence. One or more dashes ('-') within a sequence indicate insertions or deletions

1.6. Primer design

Primer programs

- http://genome-www2.stanford.edu/cgi-bin/SGD/web-primer
- http://www-genome.wi.mit.edu/cgi-bin/primer/primer3_www.cgi
- http://www.idtdna.com

Primer selection using the program at the following website: http://genome-www2.stanford.edu/cgi-bin/SGD/web-primer

- Enter DNA sequence where prompted
- Select purpose: PCR or Sequencing and then click on submit
- On the Web Primer: Parameters, select parameters according to what is required.
- The results page: If selection of primers was not successful, instructions will be given to change certain parameters e.g. Tm or GC.
- If primer selection was successful, results will be given with a best primer pair followed by a list of all other possible primers.

1.7. Other online programs

Web	Programs
http://www.firstmarket.com/cutter/cut2.html.	A free online program to help restriction map DNA sequences).
http://www.ch.embnet.org	• Basic and advanced BLAST (Sequence searches) • ClustalW (Multiple sequence alignment) • LALIGN (Alignment of 2 sequences) • FDF (Sequence searches by Smith-Waterman) • TMPRED (Transmembrane regions detection) • PRSS (Optimal score of an alignment). • COILS (Coiled domains prediction). • BOXSHADE (Multiple alignments designer) T-COFFEE (Multiple sequence alignment) PCR (Virtual PCR tool
Fetch	(Get sequences
http://www.highveld.com/pcr.html	• Polymerase Chain Reaction (PCR) jump station • Information and links on all aspects of PCR • Comprehensive directory for PCR and PCR related techniques and protocols
http://www.justbio.com	• Need to register before use. (Free to use) • DNA/RNA/Protein Cleaner (With this tool you can simply and quickly clean your sequences from unwanted text marks and spaces) • Translator (translate your DNA sequences from 1 to 6 phases in formated outputs) • NetPlasmid (Draw coloured plasmid maps) • ORFviewer (Find ORF regions on your DNA sequence. You get a graphical output on the 6 phases [direct and reverse orientations]) • Aligner (Align multiple sequences DNA or protein, based on a Clustalw engine) • Cutter (get the list of all enzyme restriction sites on your sequence) • Complementor (Get the complementary [reversed or not] of your DNA sequence) • EnzFinder (Search for commercial data

		about any Enzyme: availability, cut sequence, references, etc) • Oligo Calc (Simply and quickly calculate your primer parameters [Tm, %GC,]) • Primer3 (Pick primers from a DNA sequence. This is a very complete tool that may be of complex usage) • PatSearch (Search PROSITE patterns on your protein sequence) • SeqPainter (Color your amino-acids sequence according to their properties [default by hydrophobicity]). • ProCalc (With this tool you can simply and quickly determine statistical, theoretical MW, pI and extinction data from your protein sequence)
5)	http://genome-www2.stanford.edu/cgi-bin/SGD/web-primer	Primer selection
6)	http://www.ccsi.com/firstmarket/firstmarket/cutter/cut2.html	Restriction Enzyme mapping
7)	http://www.genebee.msu.su/services/rna2_full.html	Predict RNA secondary structure
8)	http://www.ebi.ac.uk/contrib/tommaso/translate.html	Translation
9)	http://www.ncbi.nlm.nih.gov/gorf/gorf.html	Identify coding regions
10)	http://motif.genome.ad.jp/	Identify functional regions- motifs (secondary structure prediction eg. Alpha helices and beta sheets)
11)	http://bimas.dcrt.nih.gov/molbio/proscan/	Predict promotor regions (putative eukaryotic Pol II promoter sequences)
12)	http://www.expasy.org/tools/tagident.html	Plot the charge of a peptide and determine isoelectric point
13)	http://www.mips.biochem.mpg.de/desc/protfam/	Identify similar proteins (protein families)
		Amino acid multiple alignments (see above)
14)	http://www.embl-idelberg.de/predictprotein/predictprotein.html	Predict protein secondary structure (full-length sequence)

1.8. Other links

RATING

*	**AVERAGE**
**	**GOOD**
***	**VERY GOOD**

WEB ADDRESS	DESCRIPTION	RATING
TOOLS		
http://tools.neb.com/NEBcutter/index.php3	restriction map DNA sequences	unknown
http://www.firstmarket.com/cutter/cut2.html	restriction map DNA sequences	**
http://darwin.bio.geneseo.edu/%7Eyin/WebGene/RE.html	restriction map DNA sequences	**
www.ch.embnet.org	online molecular tools	***
http://www.justbio.com/	online molecular tools	***
http://evolution.genetics.washington.edu/phylip/software.html	phylogeny programs	***
http://www-genome.wi.mit.edu/cgi-bin/primer/primer3_www.cgi	primer program	**
http://eatworms.swmed.edu/~tim/primerfinder/	primer program	unknown
http://genome-www2.stanford.edu/cgi-bin/SGD/web-primer	primer program	***
http://alces.med.umn.edu/rawtm.html	Tm Determination	*
http://mbcf.dfci.harvard.edu/docs/oligocalc.html	oligonucleotide calculator	**
http://www.biocalc.com/biochem.html	Biomedical Calculators for Biochemistry/DNA	**
http://arbl.cvmbs.colostate.edu/molkit/rtranslate/	reverse translate a protein	**
http://smart.embl-heidelberg.de/	search for domains within a protein	***
http://genome.jouy.inra.fr/cgi-bin/CloneIt/CloneIt	online program finding subcloning strategies	unknown
SEARCH ENGINES		
http://www.nln.com/	general search engine	*
www.scirus.com	scientific search engine	**
www.highway61.com	general search engine	**
www.google.com	general search engine	***
JOURNALS		
www.pubmedcentral.nih.gov	life sciences journals	**
Http://bimol.unmc.edu/journals.html	journals online	**
OTHER		
http://www.nbif.org/links/1.12.php	links to biotechnology	*
http://www.sacs.ucsf.edu/Links/BioInfo_Meta.html	Molecular Biology and BioInformatics Websites	*
www.highveld.com	Internet Directory of Molecular Biology and Biotechnology	**
www.cellbio.com	Informational Resource for Cell and Molecular Biologists	**
http://www.up.univ-mrs.fr/~wabim/english/logligne.html	list of and links to online analysis tools	**
http://www.nes.cov.ac.uk/mullenger/biores/Bioinformatics.htm	bioinformatics links	**
http://biology.fullerton.edu/biol402/molbio.html	Molecular links	**
http://www.che.utexas.edu/georgiou/Info/links.htm	tools and links to molecular biology and other	**
http://restools.sdsc.edu/	list of electronic and internet accessible tools	***
www.bioexplorer.net	bioscience links	***
http://www.biol.net/	links to the biomedical research community	***
http://www.biochem.ucl.ac.uk/~mar/bookmark.htm	online bioinformatics	***
http://molbiol-tools.ca/	online analysis tools	***

1.9. Things to do with a DNA sequence

- **Database searching** (pairwise alignments to sequences in public databases) BLAST
- **identify closest relative** in public sequence databanks
- **identify similar sequences**
  http://www.ncbi.nlm.nih.gov/blast
- **Multiple alignments** ClustalW
- **To identify conserved and non-conserved regions for primer design**
  - Group-specific Primer design
  - Pathotyping primer design

1.10. Phylogenetic analysis

http://clustalw.genome.ad.jp/
http://sdmc.krdl.org.sg:8080/~lxzhang/phylip/

Phylogenetics *is that field of biology which deals with identifying and understanding the relationships between the many different kinds of life on earth*

Epidemiology *is the study of disease in the population, defining its incidence and prevalence, examining the role of external influences such as diet or toxic substances, and examining appropriate or curative measures*

Molecular epidemiology*: determining the phylogenetic relationships of viruses or bacteria (by examining their nucleic acid sequences) in a host population (e.g. animals or plants), in order to describe the incidence and spread of the disease, and to trace the source*

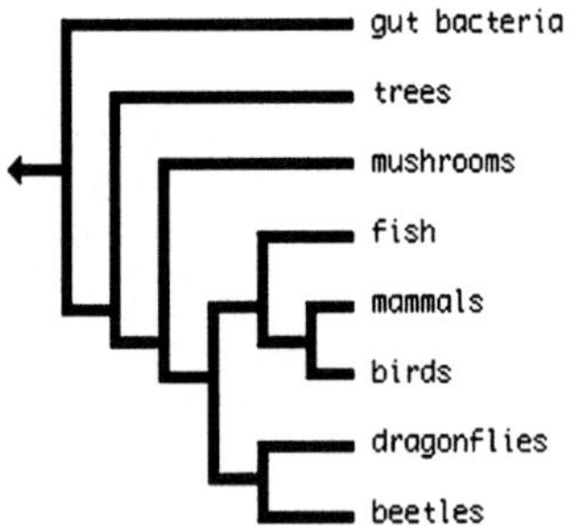

Example of phlogenetic tree

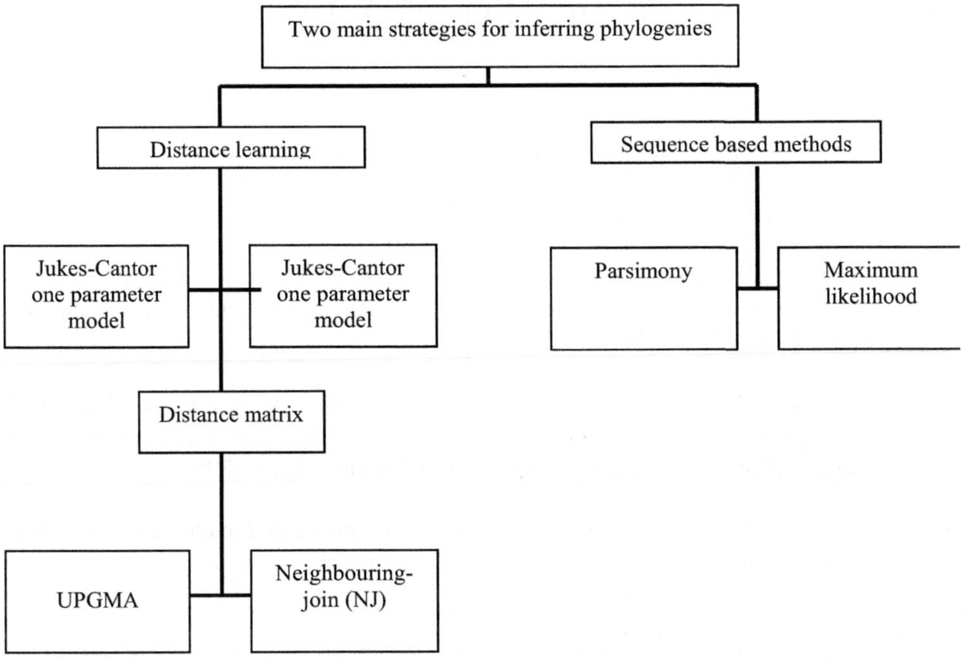

- Distance based methods are quick and easy and useful for most superficial epidemiological phylogenies. More serious phylogenetic studies (evolutionary) usually require extensive application of sequence based methods

- Boot-strapping is performed to test robustness of a phylogenetic arrangement. This is done by random sampling of nucleotides from a multiple alignment to build a new tree. A value of 80-100% means a strong support for branching order, whereas values below 70% indicate that a grouping in non-significant. Bootstrapping is usually done with distance and parsimony methods

ORIGINAL PAPERS

1. Akane. C.E. 1994. Identification of the heme compound co-purified with deoxyribonucleic acid (DNA) from blood strains, a major inhibitor of PCR. *J. of Forensic Science*, 39, 362-372

2. Allen, R.C., Graves, G. & Budowle, B. 1989. Polymerase chain reaction amplification products separated on rehydratable polyacrylamide gels and stained with silver. B*ioTechnique,* 7, 736-744

3. Armstrong, R.M., Samuel, A.R., Carpenter, W.C., Kant, R. & Knowles, N.J. 1994. A comparative study of serological and biochemical methods for strain differentiation of foot-and-mouth disease type A viruses. *Vet. Microbiol.*,39, 285-298

4. Bagnarelli, P., Menzo, A. & Valenza, A. 1995. Quantitative molecular monitoring of human immunodeficiency virus type activity during therapy with specific antiretroviral compounds. *J. of Clin Microbiol.*, 33, 16-23

5. Ballagai-Pordány, A. & Belák, S. 1996. The use of mimics as internal standards to avoid false negatives in diagnostic PCR. *Mol. Cell. Probes*, 10 (3), 159-164

6. Bassler, H. 1995. Use of a fluorogenic probe in a PCR based assay for the detection of Listeria monocytogenes. *Applied and Environmental Microbiology*, 61 (10), 3724-3728

7. Batzer, M.A., Carlton, J.E. & Deininger, P.L. 1991. Enhanced evolutionary PCR using oligonucleotides with inosine at the 3'terminus. *Nucl Acids Res* 19, 5080

8. Belák, S. & Ballagi-Pordány, A. 1993. Application of the polymerase chain reaction (PCR) in veterinary diagnostic virology. *Vet Res Commun.*, 17 (1), 55-72

9. Belák, S. & Ballagi-Pordány, A. 1993. Experiences on the application of the polymerase chain reaction in a diagnostic laboratory. *Mol Cell Probes*, 7 (3), 241-248

10. Bell, D.A. & De-Marini, D.M. 1991. Excessive cycling converts PCR products to random-length higher molecular weight fragments. *Nucleic Acids Res.*, 19, 5079

11. Bernet, C., Carret, M., de Barbeyrac, B., Bebear, C. & Bonnet, J. 1989. Detection of Mycoplasma pneumoniae by using the polymerase chain reaction. *J Clin Microbiol.,* 27, 2492-296

12. Bisno, A.L., Gerber, M.A., Gwaltney, J.M., Kaplan, E.L. & Schwartz, R.H. 1997. Diagnosis and management of group A streptococcal pharyngitis: a practice guideline. *Clin Infect Dis.*, 25,574-383

13. Brandt, M.E., Padhye, A.A., Mayer, L.W. & Holloway, B.P. 1998. Utility of random amplified polymorphic DNA PCR and TaqMan automated detection in molecular identification of Aspergillus fumigatus. *J Clin Microbiol.*, 36, 2057-2062

14. Breslauer, K.J., Frank, R., Blbcker, H. & Marky, L.A. 1986. Predicting DNA duplex stability from the base sequence. *Proc Natl Acad Sci USA*, 83, 3746-3750

15. Buchbinder, A., Jospehs, S.F., Ablashi, D., Salahuddin, S.Z., Klotman, M.E., Manak, M., Krueger, G.R., Wong-Staal, F. & Gallo, R.C. 1988. Polymerase chain reaction amplification and in situ hybridisation for the detection of human B-lymphotropic virus. *J Virol Methods*, 21, 191-197

16. Caetano-Anolles, G., Bassam, B.J. & Gresshoff, P.M. 1991. DNA amplification fingerprinting using very short arbitrary oligonucleotide primers. *BioTechnology,* 9, 553-557

17. Carman, W.F., Jacyna, M.R., Hadzizyannis, S., Karayinnis, P., McGravey, M.J., Makris, A. & Thomas, H.C. 1989. Mutation preventing formation of hepatitis B e antigen in patients with chronic hepatitis B infection. *Lancet II*, 588-591

18. Carman, W.F. & Kidd, A.H. 1989. An assessment of optimal conditions for amplification of HIV cDNA using Thermus aquaticus polymerase. *J Virol Methods,* 23, 277-289

19. Chamberlain, R.W., Wamwayi, H.M., Hockley, E., Shaila, M.S., Goatley, L., Knowles, N.J. & Barrett, T. 1993. Evidence for different lineages of rinderpest virus reflecting their geographic isolation. *J Gen Virol.*, 74, 2775-2780

20. Cheng S. 1994. Long PCR. *Nature*, 369, 684-685

21. Cheng S. 1995. Template Integrity is Essential for PCR Amplification of 20- to 30-kb Sequences from Genomic DNA. *PCR Methods and Applications*, 4 (5), 294-298

22. Cheng, S. 1994. Effective amplification of long targets from cloned inserts and human genomic DNA. *Proc Natl Acad Sci USA,* 91, 5695-5699

23. Chou, Q., Russell, M., Birch, D.E., Raymond, J. & Bloch, W. 1992. Prevention of pre-PCR mis-priming and primer dimerization improves low-copy-number amplifications. *Nucl Acids Res.*, 20, 1717-1723

24. Cole, S.A. & J.E. Hixson. 1995. Baboon lipoprotein lipase: cDNA sequence and variable tissue-specific expression of two transcripts. *Gene*, 161, 265-269

25. D'Aquila, R.T., Bechtel, L.J., Videler, J.A., Eron, J.J., Gorczyca, P. Kaplan, J.C. 1991. Maximizing sensitivity and specificity of PCR preamplification heating. *Nucl Acids Res.,* 19, 3749

26. Dedieu, L., Mady, V. & LeFevre, P.C. 1994. Development of a selective polymerase chain reaction assay for the detection of *Mycoplasma mycoides* subsp. *mycoides* S.C. (contagious bovine pleuropneumonia agent). V*et. Microbiol*., 42, 327-339

27. Demmler, G.J., Buffone, G.J., Schimbor, C.M. & May, R.A. 1988. Detection of cytomegalovirus in urine from newborns by using polymerase chain reaction DNA amplification. *J Infect Dis.*, 158, 1177-1184

28. DiCesare, J. 1993. A High-Sensitivity Electrochemiluminescence-Based Detection System for Automated PCR Product Quantitation. *BioTechniques,* 15 (1), 152-156

29. Dieffenbach, C. 1993. Setting up a PCR laboratory. PCR Methods and Applications, 3,S2-S7

30. Dohner, D. 1995. Inhibition of PCR by mireral oil exposed to UV irradiation for prolonged periods. BioTechniques, 18 (6), 964-967

31. Donn, A., Martin, L.A. & Donaldson, A.I. 1994. Improved detection of persistent foot-and-mouth disease infection in cattle by the polymerase chain reaction. *J. Virol. Meth.*, 49, 179-186

32. Dragon., E. 1993. Handling Reagents in the PCR Laboratory. *PCR Methods and Applications,* 3, S8-S9

33. Dunn, C.S. & Donaldson, A.I. 1997. Natural adaptation to pigs of a Taiwanese isolate of foot-and-mouth disease virus. V*et. Rec.*, 141, 174-175

34. Eckert, K.A. & Kunkel, T.A. 1991. The fidelity of DNA polymerases used in the polymerase chain reactions. In PCR A Practical Approach; Eds. Mcpherson MJ, Quirke P, Taylor GR, *IRL Press*

35. Edwards, M. 1994. Multiplex PCR: advantages, development, and applications. *PCR Methods and Applications*, 3, S65-S71

36. Erlich, H.A., Gelfand, D. & Sninsky, J.J. 1991. Recent advances in the polymerase chain reaction. *Science*, 252, 1643-1651

37. Espy, M.J. & Smith, T.F. 2000. Detection of cytomegalovirus (CMV) in clinical specimens by LightCycler PCR (abstract). Abstracts of the 100th General Meeting of the American Society for Microbiology, No. C-62. Los Angeles, CA, May 21-25, 2000

38. Espy, M.J., Patel, R., Paya, C.V. & Smith, T.F. 2001. Quantification of Epstein-Barr virus (EBV) viral load in transplant patients by LightCycler PCR (abstract). Abstracts of the 101st General Meeting of the American Society for Microbiology, No. C- 148. Orlando, Florida, May 20-24, 2001

39. Espy, M.J., Rys, P.N., Wold, A.D., Uhl, J.R., Sloan, L.M., Jenkins, G.D., Ilstrup, D.M., Cockerill, F.R. III, Patel, R., Rosenblatt, J.E. & Smith, T.F. 2001. Detection of herpes simplex virus DNA in genital and dermal specimens by LightCycler PCR after extraction by IsoQuick, MagNA Pure, and BioRobot 9604. *J Clin Microbiol*, 39, 2233-2236

40. Ferre, F. 1994. Quantitative PCR: An Overview. The Polymerase Chain Reaction. Ed. Mullis. Boston: Birkhauser

41. Ferre, F. 1992. Quantitative or Semi-quantitave PCR: Reality Versus Myth. *PCR Methods and Applications*, 2 (1), 1-9

42. Finckh. U., Lingenfelter, P.A. & Myerson, D. 1991. Producing single-stranded DNA probes with the *Taq* DNA polymerase: A high yield protocol. *BioTechniques*, 10, 35-39

43. Foord, O. 1994. Long-distance PCR. *PCR Methods and Applications,* 3 (6), S149-S161

44. Forsyth, M.A. & Barrett, T. 1995. Evaluation of a polymerase chain reaction for the detection and characterisation of rinderpest and peste des petits ruminants for epidemiological studies. *Virus Res.*, 39, 151-163

45. Frohman. M. 1994. On beyond classic RACE (rapid amplification of cDNA ends). *PCR Methods and Applications*, 4 (1), S40-S58

46. Frye, R.A., Benz, C.C. & Liu, E. 1989. Detection of amplified oncogenes by differential polymerase chain reaction. *Oncogene*, 4, 101-105

47. Fujiwara, H. 1995. PCR with Deoxyinosine-containing Primers Using DNA Polymerases with Proofreading Activity. *PCR Methods and Applications*, 4 (4), 239-240

48. Gelfand, D.H. Taq DNA polymerase. 1989. In PCR Technology Principles and Applications for DNA Amplification; Ed. Erlich H.A., *Stockton Press*

49. Gelfand, D.H. & White, T.J. 1990.Thermostable DNA polymerases. In PCR Protocols, A Guide to Methods and Applications; Eds. Innis, M.A., Gelfand, D.H., Sninsky, J.J., White, T.J., *Academic Press*

50. Gibbs, R.A., Nguyen, P.N., McBride, L.J., Koepf, S.M. & Caskey, C.T. 1989 Identification of mutations leading to the Lesch-Nyhan syndrome by automated direct DNA sequencing of in vitro amplified cDNA. *Proc Natl Acad Sci USA,* 86, 1919-1923

51. Gibbs, R.A., Ngyen, P.N. & Caskey, C.T. 1989. Detection of single DNA base differences by competitive oligonucleotide priming. *Nucl Acids Res.,* 17, 2437-2448

52. Gibson, U.E.M., Heid, C.A. & Williams, P.M. 1996. A novel method for real-time quantitative RT-PCR. *Genome Res.*, 6, 995-1001

53. Gilliland, G., Perrin, S., Blanchard, K. & Bunn, H.F. 1990. Analysis of cytokine mRNA and DNA: detection and quantitation by competitive polymerase chain reaction. *Proc Natl Acad Sci USA*, 87, 2725-2729

54. Greenberg, S.J., Ehrlich, G.D., Abbot, M.A., Hurwitz, B.J., Waldman, T.A. & Poiesz, BJ. 1989. Detection of sequences homologous to human retroviral DNA in multiple sclerosis by gene amplification. *Proc Natl Acad Sci USA*, 86, 2878-2882

55. Gyllensten, U.B. & Erlich, H.A. 1988. Generation of single stranded DNA by the polymerase chain reaction and its application to direct sequencing of the HLA-DQA locus. *Proc Natl Acad Sci USA*, 85, 7652-7657

56. Haff, L. 1994. Improved quantitative PCR using nested primers. *PCR Methods and Applications*, 3 (6), 332-337

57. Hanni, C. 1995. Isopropanol precipitation removes PCR inhibitors from ancient bone extracts. *Nucleic Acids Research, 23 (5), 881-882*

58. Harding, M.J., Prud'homme, I., Gradil, C.M., Heckert, R.A., Riva, J., McLaurin, R., Dulac, G.C. & Vydelingum, S. 1996. Evaluation of nucleic acid amplification methods for the detection of hog cholera virus. *J. Vet. Diagn. Invest.*, 8, 414-419

59. Hartley, J. 1993. Dealing with Contamination: Enzymatic Control of Carryover Contamination in PCR. *PCR Methods and Applications,* 3 (2), S10-S14

60. Heid, C.A., Stevens, J., Livak, K.J. & Williams, P.M. 1996. Real-time quantitative PCR. *Genome Res.* 6, 986-994

61. Her, C. 1995. Rapid Restriction Mapping by use of long PCR. *BioTechniques*, 19 (4), 530-532

62. Heroux, J. 1995. Quantitative Analysis of Specific mRNA Transcripts Using a Competitive PCR assay with Electrochemiluminiscent Detection. *PCR Methods and Applications*, 4 (6), 327-330

63. Holland, P.M., Abramson, R.D., Watson, R. & Gelfand, D.H. 1991. Detection of specific polymerase chain reaction product by utilizing the 5'>3' exonuclease activity of *Thermus aquaticus* DNA polymerase. *Proc Natl Acad Sci USA,* 88, 7276-7280

64. Holland, P. 1991. Detection of specific polymerase chain reaction product by utilizing the 5'->3' exonuclease activity of Thermus aquaticus DNA polymerase. *Proc Natl Acad Sci USA*, 88, 7276-7280

65. Hoof van Huijsduijnen, R.A.M., Ayala, G. & DeLamarter, F. 1992. A means to reduce the complexity of oligonucleotides encoding degenerate peptides. *Nucl Acids Res.*, 20, 919

66. Horn, G.T., Richards, B. & Klinger, K.W. 1989. Amplification of highly polymorphic VNTR segment by the polymerase chain reaction. *Nucl Acids Res.,* 17, 2140

67. Huang, S. 1994. Long-range jumping of incompletely extended polymerase chain fragments generates unexpected products. *BioTechniques*, 16 (2) 242-246

68. Innis, M.A., Gelfand, D.H. & Sninsky, J.J. (eds), 1995. PCR Strategies. Academic Press, Inc, California, USA

69. Innis, M.A., Mayambo, K.B., Gelfand, D.H. & Brow, M.A.D. 1988. DNA sequencing with *Thermus aquaticus DNA* polymerase and direct sequencing of polymerase chain reaction-amplified DNA. *Proc Natl Acad Sci USA*, 85, 9436-9440

70. Ireland, D.C. & Binepal, Y.S. 1998. Improved detection of capripoxvirus in biopsy samples by PCR. *J. Virol. Meth*., 74, 1-7

71. Ishaq, M. 1994. Differential expression of mRNAs for JC virus large and small tumor antigens in brain tissues from progressive multifocal leukoencephalopathy patients with and without AIDS. *PNAS,* 91, 8283-8287

72. Jeffreys, A.J., MacLeod, A., Tamaki, K., Neil, D. & Monckton, D.G. 1991. Minisatellite repeat coding as a digital approach to DNA typing. *Nature*, 354, 204-209

73. Jeffreys, A.J., Neumann, R. & Wilson, V. 1990. Repeat unit sequence variation in minisatellites: a novel source of DNA polymorphism for studying variation and mutation by single molecule analysis. *Cell*, 60, 473-485

74. Jessen-Eller, K.1994. Quantitation of metallothionein mRNA by RT-PCR an chemiluminescence. *BioTechniques, 17 (5), 962-972*

75. Jin, C. 1994. Rapid construction of deleted DNA fragments for use as internal standards in competitive PCR. *PCR Methods an Applications*, 3, 252-255

76. Jones, D.H. & Winistorfer, C. 1992. Sequence specific generation of a DNA panhandle permits PCR amplification of unknown flanking DNA. *Nucl Acids Res.*, 20, 595-600

77. King, A.M.Q., Underwood, B.O., McCahon, D., Newman, J.W.I. & Brown, F. 1981. Biochemical identification of viruses causing the 1981 outbreaks of foot and mouth disease in the UK. *Nature*, 293, 479-480

78. Koike, K., Urata, Y. & Koike, M. Human Pyruvate Dehydrogenase beta-subunit-gene, complete cds. Un-published data, EMB L accession number D90086

79. Kwok, S. 1991. A Guide to the Design and Use of Mismatched and Degenerate Primers. *PCR Methods* and Applications, 3 (4), S39-S47

80. Kwok, S., Kellogg, D.E., McKinney, N., Spasic, D., Goda, L., Levenson, C. & Sninsky, J.J. 1990. Effects of primer-template mismatches on the polymerase chain reaction: human immunodeficiency virus type I model studies. *Nucl Acids Res.*, 18, 999-1005

81. Lawyer, F.C., Stoffel, S., Saiki, R.K., Mayambo, K., Drummond, R. & Gelfand, D.H. 1989. Isolation, characterization, and expression in *Escherichia coli* of the DNA polymerase gene from *Thermus aquaticus. J Biol Chem.*, 246, 6427-6437

82. Lee,O.C. 1993. Allelic discrimination by nick-translation PCR with fluorogenic probes. *Nucleic Acids Research*, 21 (6), 3761-3766

83. Lee, O.C. 1995. Improved direct PCR screen for bacterial colonies: wooden toothpicks inhibit PCR amplification. *BioTechniques*, 18 (2), 225-226

84. Livak., K. 1995. Towars fully automated genome wide polymorphism screening. *Nature Genetics*, 9, 341-342

85. Livak, K. 1995. Olignucleotides with Fluorescent Dyes at Opposite Ends Provide a Quenched Probe System Useful for Detecting PCR Product and Nucleic Acid Hybridisation. *PCR Methods and Applications*, 4 (6), 357-362

86. Livak, K.J., Flood, S.J.A., Marmaro, J., Giusti, W. & Deetz K. 1995. Oligonucleotides with fluorescent dyes at opposite ends provide a quenched probe system useful for detecting PCR product and nucleic acid hybridisation. *PCR Methods Appl.*, 4, 357-362

87. Lo, Y.M., Mehal, W.Z. & Fleming, K.A. 1989. In vitro amplification of hepatitis B virus sequences from liver tumour DNA and from paraffin wax embedded tissues using the polymerase chain reaction. *J Clin Pathol.*, 42, 840-846

88. Lockey, C., Otto, E. & Long, Z. 1998. Real-time fluoresence detection of a single DNA molecule. *Biotechniques*. 24, 744-746

89. Loeffelholz, M.J., Thompson, C.J., Long, K.S. & Gilchrist, M.J.R. 1999. Comparison of PCR, culture, and direct fluorescent-antibody testing for detection of *Bordetella pertussis. J Clin Microbiol.*, 37,2872-2876

90. Longo, P. 1990. Use of uracil DNA glycosylase to control carry-over contamination in polymerase chain reactions. *Gene*, 93,125-128

91. Mazars, G.R., Moyret, C., Jeanteur, P. & Theillet, C.G. 1991. Direct sequencing by thermal asymmetric PCR. *Nucl Acids Res.*, 19, 4783

92. McGoldrick, A., Lowings, J.P., Ibata, G., Sands, J.J., Belak, S. & Paton, D.J. 1998. A novel approach to the detection of classical swine fever virus by RT-PCR with a fluorogenic probe (TaqMan). *J Virol Methods*, 72, 125-135

93. McPherson, P. PCR with highly degenerate primers. *PCR: A Practical approach.* Ed. McPherson, P.. 1991. New York: Oxford University pp.171-186

94. Menon, R. 1991. RT-PCR artifacts from processed pseudogenes. *PCR Methods and Applications*, 1, 70-71

95. Mertz, C.E. 1994. Nucleotide imbalance and PCR: Effects on DNA amplification and synthesis of high specific activity radiolabeled DNA probes. *Anal. Biochem.*, 221, 160-165

96. Meyer, R.F., Brown, C.C., House, C., House, J.A. & Molitor, T.W. 1991. Rapid and sensitive detection of foot and-mouth disease virus in tissues by enzymatic RNA amplification of the polymerase gene. *J. Virolog. Meth.*, 34, 161-172

97. Mezei, L.M. 1990. Effect of oil overlay on PCR amplification. *Amplifications A Forum for PCR Users*, 4, 11-13

98. Mullis, K.B. 1991. The polymerase chain reaction in an anemic mode: How to avoid cold oligodeoxyribonuclear fusion. *PCR Methods and Applications*, 1, 1-4

99. Mullis, K.B. & Faloona, F.A. 1987. Specific synthesis of DNA in *vitro* via a polymerase-catalyzed chain reaction. *Methods Enzymol.*, 155, 335-350

100. Murphy, G.F. 1993. Rethenium Complexes as Luminescent Reporters of DNA. Methods in Enzymology. Anonymous Academic Press, 226pp. 576-595

101. Murray, P.R. & Washington, J.A. II. 1975. Microscopic and bacteriologic analysis of expectorated sputum. *Mayo Clin Proc*, 50, 339-344

102. Myers, R. 1995. Amplification of RNA: High-Temperature Reverse Transcription and DNA Amplification *with Thermus Thermophilus DNA Polymerase. PCR Strategies. Ed. Innis. San Diego: Academic Press. Inc.* p 58-68

103. Odelberg, S.J., R.B. Weiss, A. & Hata, E. 1995. Template switching during DNA synthesis by Thermus aquaticus DNA polymerase I. *Nucleic Acids Research*, 23, 2049-2057

104. Osterrieder, N., Hubert, P.H. & Brandmuller, C. 1994. A touchdown PCR for the differentiation of equine herpesvirus type 1 (EHV-1) field strains from the modified live vaccine strain Rach. *J of Virol Methods,* 50, 129-136

105. Palmenberg, A.C. 1989. Sequence alignments of picornaviral capsid proteins. In: Molecular Aspects of Picornaviral Infection and Detection. Semler, B.L. and Ehrenfeld, E. (eds.). American Society for Microbiology, Washington, USA. pp. 211-241

106. Perrin, S. & Filliland, G. 1990. Site-specific mutagenesis using asymmetric polymerase chain reaction and a single mutant primer. *Nucl Acids Res.*, 18, 7433-7438

107. Peterson, R. 1993. Cross-Species Polymerase Chain Reaction: Cloning of TATA Box-Binding Proteins. *Methods in Enzymology: Anonymous Academic Press,* pp. 493-507.

108. Prince, S.E. 1992. PCR: How to Kill Unwanted DNA. *BioTechniques,* 12, 358-360

109. Rao. V.B. 1994. Direct Sequencing of Polymerase Chain Reaction-Amplified DNA. *Anal Biochem,* 216, 1-14

110. Rappolee, D.A. 1990. Optimizing the sensitivity of RT-PCR. *Amplifications A Forum for PCR Users*, 4, 5-7

111. Reischl, U., Wittwer, C. & Cockerill, F. (eds), 2002. Rapid cycle real-time PCR. Methods and applications. Springer-Verlag Berlin Heidelberg, Germany.

112. Rolfs, A., Schuller, I., Finckh, U. & Weber-Rolfs, I. (eds), 1992. PCR: Clinical Diagnostics and research. Springer-Verlag Berlin Heidelberg, Germany.

113. Roux, K. 1994. Using mismarched Primer-Template Pairs in Touchdown PCR. *BioTechniques,* 16 (5), 812-814

114. Roux, K. 1995. Optimization and troubleshooting in PCR. PCR Methods and Applications. 4 (5), S185-S194

115. Ruano, G., Brash, D.E. & Kidd, K.K. 1991. PCR: The first few cycles. *Amplifications, A Forum for PCR Users,* 7, 1-4

116. Ruano, G., Fenton, W. & Kidd, K.K. 1989. Biphasic amplification of very dilute DNA samples via "booster" PCR. *Nucl Acids Res.*, 17, 5407

117. Ruster, D.W. 1995. Quantification of Hepatitis C Virus RNA by Competitive Reverse Transcription and Polymerase Chain Reaction Using a Modified Hepatitis C Virus RNA Transcript. *Anal. Biochem.*, 224, 597-600

118. Rychlik, W. 1995. Priming efficiency in PCR. *BioTechniques*, 18 (1), 84-90

119. Rychlik, W., Spencer, W.J. & Rhoads, R.E. 1990. Optimization of the annealing temperature for DNA amplification in *vitro. Nucl Acids Res.*, 18, 6409-6412

120. Rychlik, W. & Rhoads, R.E. 1989. A computer program for choosing optimal oligonucleotides for filter hybridisation, sequencing and in *vitro* amplification of DNA. *Nucl Acids Res.*, 17, 8543-8551

121. Saiki, R.K., Gelfand, D.H., Stoffel, S, Scharf, S.J., Higuchi, R., Horn, G.T., Mullis, K.B. & Erlich, H.A. 1988. Primer-directed enzymatic amplification of DNA with a thermostable DNA polymerase. *Science*, 239, 487-491

122. Saiki, R.K., Gelfand, D.H., Stoffel, S., Scharf, S.J., Higuchi, R., Horn, G.T., Mullis, K.B. & Erlich, H.A. 1998. Primer-directed enzymatic amplification of DNA with a thermostable DNA polymerase. *Science*, 239, 487-491

123. Saiki, R.K., Scharf, S., Faloona, F., Mullis, K.B., Horn, G.T., Erlich, H.A. & Arnheim, N. 1985. Enzymatic amplification of ß-globin genomic sequences and restriction site analysis for diagnosis of sickle cell anemia. *Science*, 230, 1350-1354

124. Shannon, A.D., Morrissy, C., Mackintosh, S.G. & Westbury, H.A. 1993. Detection of hog cholera virus antigens in experimentally-infected pigs using an antigen-capture ELISA. *Vet. Microbiol.,* 34, 233-248

125. Shibata, D., Martin, W.J., Appleman, M.D., Causey, D.M., Leedom, J.M. & Arnheim, N. 1988. Detection of cytomegalovirus DNA in peripheral blood of patients infected with human immunodeficiency virus. *J Infect Dis.*, 158, 1185-1192

126. Sommer, R. & Tautz, D. 1989. Minimal homology requirements for PCR primers. *Nucl Acids Res*, 16, 6749

127. Taylor, G.R. 1991. Polymerase chain reaction: basic principles and automation. In PCR A Practical Approach; Eds.. Mcpherson M.J., Quirke P. & Taylor G.R. (Eds). *IRL Press*

128. Thein, S.L. & Wallace, R.B. 1986. The use of synthetic oligonucleotides as specific hybridisation probes in the diagnosis of genetic disorders. In Human genetic diseases, a practical approach. Eds. Davis K.E. *IRL Press, Herndon, Virginia*

129. Tindall, K.R. & Kunkel, T.A. 1988. Fidelity of DNA synthesis by *the Thermus aquaticus* DNA polymerase. *Biochemistry*, 27, 6008-6013

130. Trower, M.K., D. Burt, I.J. & Purvis, T. 1995. Fluorecent dyeprimer cycle sequencing using unpurified PCR products as templates : development of a protocol amenable to highthroughput DNA Sequencing. *Nucleic Acids Research,* 23, 2348-2349

131. Tyagi, S. & Kramer, F.R. 1996. Molecular beacons: probes that fluoresce upon hybridisation. *Nature Biotechnol.*, 14 (3), 303-308

132. Uhl, J.R., Sandhu, G.S., Kline, B.C. & Cockerill, F.R. III. 1996. PCR-RFLP detection of point mutations in the catalase-peroxidase gene *(katG) of Mycobacterium tuberculosis* associated with isoniazid resistance. In: Persing DH (ed) PCR Protocols for emerging infectious diseases. *American Society for Microbiology*, Washington, DC

133. Van Devanter, D.R., Warrener, P. & Bennett, L. 1996. Detection and analysis of diverse herpesviral species by consensus primer PCR. *J Clin Microbiol.*, 34 (7), 1666-1671

134. Vet, J.A., Majithia, A.R. & Marras, S.A. 1999. Multiplex detection of four pathogenic retroviruses using molecular beacons. *Proc Natl Acad Sci. USA* 96 (11), 6394-6399

135. Walsh, P. 1992. Preferential PCR Amplification of Alleles: Mechanisms and Solutions. *PCR Methods an Applications,* 1 (4), 241-250

136. Wayne, P.A. November 1998. Basic cost accounting for clinical services; approved guidelines. National Committee for Clinical Laboratory Standards, GP11-A, Vol. 18,

137. White, B.A. (ed), 1993. PCR Protocols. In Methods in Molecular Biology, volume 15. Humana Press, New Jersey, USA

138. Wilkinson, E. 1995. Development of Competitive PCR and the QPCR System 5000 as a Transcription-based Screen. *PCR Methods and Applications*, 4 (6), 363-367

139. Wilkenson, E.T., S. Cheifetz, & S.A. Degrandis. 1995. Development of competitive PCR and the QPCR system 5000 as a transcription-based screen. *PCR Methods and Applications*, 4, 363-367

140. Williams, J.G.K., Kubelik, A.R., Livak, K.J., Rafalski, J.A. & Tingey, S.V. 1990. DNA polymorphisms amplified by arbitrary primers are useful as genetic markers. Nucl Acids Res., 18, 6531-6535

141. Wilson, R.K., Chen, C. & Hood, L. 1990. Optimization of asymmetric polymerase chain reaction for rapid fluorescent DNA sequencing. *BioTechniques*, 8, 184-189

142. Wu, C.H., C.C. Wang, H.M. & Yang, E. 1995. Sequences at four termini of the giardiavirus double-stranded RNA. *Gene*, 158, 129-131

143. Yap, E.P.H. & McGee, O'D. 1991. Short PCR product yields improved by lower denaturation temperatures. *Nucl Acids Res.*, 19, 1713

Books and reviews

1. Bernd Kochanowski and Udo Reischl (Eds). Quantitative PCR Protocols: Methods in molecular medicine. 1999 Humana Press, Totowa, New Jersey

2. H. Erlich (Ed). PCR Technology: Principles and Applications for DNA Amplification. 1992 W.H. Freeman and Company, New York

3. J A. Scheppler, P.E. Cassin and R.M. Gambier. Biotechnology Explorations: applying the Fundamentals. 2000 ASM Press, Washington, D.C.

4. J. Mcpherson, B. D. Hames and G. R. Taylor (Eds). PCR 2: A practical Approach. 1995 IRL Press, Oxford University Press. New York

5. OIE, Manual of Standards for Diagnostic Tests And Vaccines (4th Edition), 2002

6. K. Doyle (Ed). Protocols and Applications Guide, 3rd Edition. 1996 Promega Corporation. USA

7. R. Rapley (Ed) The Nucleic Acid Protocols Handbook. 1999 Humana Press, Totowa, New Jersey

8. Sambrook, J., Fritsch, E.F. & Maniatis, T. Molecular cloning. A laboratory manual, 2nd Edition. 1998 Cold Spring Harbor

9. Steffen. C. (Ed). PCR Applications Manual, 2nd edition. 1999 Roche Diagnostics GmbH, Mannheim, Germany.

10. Y. Becker and G. Darai (Eds). PCR: Protocols for Diagnosis of Human and Animal Virus Diseases. 1995 Springer, New York

GLOSSARY

Glossary

A	
AFLP	Amplified fragment length polymorphism.
AGAROSE GEL ELECTROPHORESIS	A method for separating nucleic acids (DNA or RNA) within a gel made of agarose in a suitable buffer under the influence of an electrical field. Suitable for separation of large fragments of nucleic acid, separation is based primarily upon the size of the nucleic acid.
ALLELE	One of several alternate forms of a gene occupying a given locus on a chromosome or plasmid.
AMINO ACIDS	The 20 basic building blocks of proteins, consisting of the basic formula NH2-CHR-COOH, where "R" is the side chain that defines the amino acid: *Nonpolar side chains (hydrophobic)* G Gly Glycine A Ala Alanine V Val Valine I Ile Isoleucine L Leu Leucine F Phe Phenylalanine P Pro Proline M Met Methionine W Trp Tryptophan C Cys Cysteine *Noncharged polar side chains (hydrophilic)* S Ser Serine T Thr Threonine Y Tyr Tyrosine N Asn Asparagine Q Gln Glutamine *Acidic side chains (very polar, hydrophilic)* D Asp Aspartic Acid E Glu Glutamic Acid *Basic side chains (very polar, hydrophilic)* K Lys Lysine R Arg Arginine H His Histidine
AMPLICON	The product' of double-stranded DNA created during PCR.
AMINO TERMINUS	Refers to the NH2 end of a *peptide* chain (by custom drawn at the left of a protein sequence).
AMPLIFICATION	The production of additional copies of a chromosomal sequence, found either as intrachromosomal or extrachromosomal DNA. Also refers to the in vitro process in the *polymerase chain reaction*.
AMPLIMER	Region of DNA sequence that is amplified during a *PCR* reaction and that is defined by a pair of *PCR primers* (these *primer* pairs are sometimes called amplimers).

Glossary

AMV	Avian myeloblastosis virus RT.
ANCHOR SEQUENCE	A hydrophobic amino acid sequence that fixes a segment of a newly synthesized, *translocating* protein within the lipid bilayer membrane of the endoplasmic reticulum.
ANNEALING	The binding of complementary DNA or RNA sequences between bases via hydrogen bonding.
ANNEALING TEMPERATURE	The temperature at which a length of single-strand (heat denatured) DNA or RNA will anneal to a complementary strand. Lower temperatures may permit non-specific binding.
ANTISENSE STRAND (OR PRIMER)	Refers to the RNA or DNA strand of a *duplex* molecule that is *complementary* to that encoding a *polypeptide*. More specifically, the DNA strand that serves as *template* for the synthesis of RNA and which is complementary to it. "Antisense *oligonucleotides*" *hybridize* to *mRNA*, and are used to prime *cDNA* synthesis.
AVIDIN	A glycoprotein that binds to *biotin* with very high affinity (Kd = 10-15).
B	
BACK MUTATION	Reverse the effect of a point or frame-shift mutation that had altered a gene; thus it restores the wild-type phenotype (see REVERTANT).
BACTERIOPHAGE	A virus that infects bacteria; often simply called a phage. The phages that are most often used in molecular biology are the *E. coli* viruses lambda, M13 and T7.
BASE	Adenine (A), cytosine (C), guanine (G) and thymine (T) are the molecules composing DNA. These are bases in DNA that form two pairs of complementary molecules, hydrogen bonds can link adenine to thymine (A_T) and guanine to cytosine (G_C). If the sequence of bases matches a complementary sequence in a second strand of DNA the bonds can hold the two strands together; thus forming double-stranded.
BASE PAIR (bp)	One pair of *complementary nucleotides* within a *duplex* strand of a nucleic acid (BP) and 10^3 bp = kb) Under Watson-Crick rules, these pairs consist of one *pyrimidine* and one *purine*: i.e., C-G, A-T (DNA) or A-U (RNA). However, "noncanonical" base pairs (e.g., G-U) are common in RNA *secondary structure*.
BIOTIN	A coenzyme that is essential for carboxylation reactions (see Avidin).
BLUNT END	A terminus of a *duplex* DNA molecule that ends precisely at a base pair, with no *overhang* (unpaired nucleotide) in either strand. Some but not all *restriction endonucleases* leave blunt ends after cleaving DNA. Blunt-ended DNA can be *ligated* nonspecifically to other blunt-ended DNA molecules (compare with STICKY END).

BOX	Refers to a short nucleic acid *consensus sequence* or *motif* that is universal within kingdoms of organisms. Examples of DNA boxes are the Pribow box (TATAAT) for *RNA polymerase*, the Hogness box (TATA) that has a similar function in eukaryotic organisms, and the homeo box. RNA boxes have also been described, such as Pilipenko's Box-A motif that may be involved in ribosome binding in some viral RNAs.
BSA	Bovine serum albumin.
C	
CAP	A 7-methyl guanosine residue linked 5' to 5' through a triphosphate bridge to the 5' end of eukaryotic *mRNAs*; facilitates initiation of *translation.*
CARBOXYL TERMINUS	Refers to the COOH end of a *peptide* chain (by custom drawn at the right of a protein sequence).
cDNA	Complementary DNA. A DNA molecule that was originally copied from an RNA molecule by *reverse transcription.* The term "cDNA" is commonly used to describe double-stranded DNA that originated from a single-stranded RNA molecule, even though only one strand of the DNA is truly complementary to the RNA.
cDNA LIBRARY	A collection of *cDNA* fragments, each of which has been cloned into a separate *vector* molecule.
CHAPERONE PROTEINS	A series of proteins present in the endoplasmic reticulum that guide the proper folding of secreted proteins through a complex series of binding and release reactions.
CHROMOSOME WALKING	The sequential isolation of *clones* carrying overlapping sequences of DNA that span large regions of a chromosome. Overlapping regions of clones can be identified by *hybridisation.*
CLONE	Describes a large number of cells, viruses, or molecules that are identical and derived from a single ancestral cell, virus or molecule. The term can be used to refer to the process of isolating single cells or viruses and letting them proliferate (as in a *hybridoma* clone, which is a "biological clone"), or the process of isolating and replicating a piece of DNA by recombinant DNA techniques ("molecular clone"). The use of the word as a verb is acceptable for the former meaning, but not necessarily the latter meaning.
CIS	As used in molecular biology, an interaction between two sites that are located within the same molecule. However, a *cis*-acting protein can either be one that acts only on the molecule of DNA from which it was expressed, *or* a protein that acts on itself (e.g., self-proteolysis).

CISTRON	A nucleic acid segment corresponding to a polypeptide chain, including the relevant *translational start (initiation)* and *stop (termination) codons*.
CODON	A nucleotide triplet (sequence of three nucleotides) that specifies a specific *amino acid*, or a *translational* start or stop.
CODON BIAS	The tendency for an organism or virus to use certain codons more than others to encode a particular amino acid. An important determinant of codon bias is the guanosine-cytosine (GC) content of the genome. An organism that has a relatively low G+C content of 30% will be less likely to have a G or C at the third position of a codon (wobble position) than a A or T to specify an amino acid that can be represented by more than one codon.
COMPETENT	Bacterial cells that are capable of accepting foreign extra-chromosomal DNA. There are a variety of processes by which cells may be made competent.
CFT	Complement Fixation Test. Serological technique whereby a fixed system of sheep red blood cells, antibody to sheep red blood cells and complement (from guinea pig sera) is titrated using a heamolysis assay. The system is challenged by the addition of antigen/antibody reactions to the titrated complement. Any reaction uses (fixes) the complement and this is assessed in the pre-titrated system where a lack of heamolysis indicates extent of test antigen/antibody reaction.
CONSENSUS SEQUENCE	A linear series of nucleotides, commonly with gaps and some degeneracy that define common features of homologous sequences or recognition sites for proteins that act on or bind to nucleic acids.
CONSERVATIVE SUBSTITUTION	Nucleotide mutation that alters the amino acid sequence of the protein, but which causes the substitution of one amino acid with another that has a side chain with similar charge/polarity characteristics (see AMINO ACID). The size of the side chain may also be an important consideration. Conservative mutations are generally considered unlikely to profoundly alter the structure or function of a protein, but there are many exceptions
CONSERVED	Similar in structure or function.
CONTAMINATION	Extraneous DNA present in the sample or reaction that will be amplified and give a false positive result.

CONTIG	A series of two or more individual DNA sequence determinations that overlap. In a sequencing project the contigs get larger and larger until the gaps between the contigs are filled in.
COSMID	A genetically engineered *plasmid* containing bacteriophage lambda packaging signals and potentially very large pieces of inserted foreign DNA (up to 50 kb) that can be replicated in bacterial cells. Cosmid cloning allows for isolation of DNA fragments that are larger than those that can be cloned in conventional plasmids.
D	
DATABASE SEARCH	Once an open *reading frame* or a partial amino acid sequence has been determined, the investigator compares the sequence with others in the databases using a computer and a search algorithm. This is usually done in a protein database such as PIR or Swiss-Prot. Nucleic acid sequences are in GenBank and EMBL databases. The search algorithms most commonly used are BLAST and FASTA.
DC	Direct current.
DEGENERACY	Refers to the fact that multiple different *codons* in *mRNA* can specify the same *amino acid* in an encoded protein.
DENATURATION	With respect to nucleic acids, refers to the conversion from double-stranded to the single-stranded state, often achieved by heating or alkaline conditions. This is also called "melting" DNA. With respect to proteins, refers to the disruption of *tertiary* and *secondary* structure, often achieved by heat, detergents, chaotropes, and sulfhydryl-reducing agents.
DENATURING GEL	An *agarose* or *acrylamide gel* run under conditions that destroy *secondary* or *tertiary* protein or RNA structure. For protein, this usually means the inclusion of 2-ME (that reduces disulfide bonds between cysteine residues) and SDS and/or urea in an acrylamide gel. For RNA, this usually means the inclusion of formaldehyde or glyoxal to destroy higher ordered RNA structures. In DNA sequencing gels, urea is included to denature dsDNA to ssDNA strands. In denaturing gels, macromolecules tend to be separated on the basis of size and (to some extent) charge, while shape and oligomerization of molecules are not important. Contrast with NATIVE GEL.
DEOXYRIBONUCLEASE (DNASE	An enzyme that specifically catalyzes the hydrolysis of DNA.

DEOXYRIBONUCLEOTIDE	*Nucleotides* that are the building blocks of DNA and that lack the 2' hydroxyl moiety present in the ribonucleotides of RNA.
DEPC	Diethyl pyrocarbonate.
DIDEOXYRIBONUCLEOTIDE (ddNTP)	A *nucleotide (dATP, dTP, dCTP and dTTP)* that lacks both 3' and 2' hydroxyl groups. Such dideoxynucleotides can be added to a growing nucleic acid chain, but do not then present a 3' -OH group that can support further propagation of the nucleic acid chain. Thus such compounds are also called "chain terminators", and are useful in DNA and RNA (ATP, GTP, CTP and UTP) sequencing reactions (see DEOXYRIBO-NUCLEOTIDE).
DIDEOXY SEQUENCING	Enzymatic determination of DNA or RNA sequence by the method of Sanger and colleagues, based on the incorporation of chain terminating *dideoxynucleotides* in a growing nucleic acid strand copied by *DNA polymerase* or *reverse transcriptase* from a DNA or RNA *template*. Separate reactions include dideoxynucleotides containing A,C, G, or T bases. The reaction products represent a collection of new-labelled DNA strands of varying lengths, all terminating with a dideoxynucleotide at the 3' end (at the site of a complementary base in the template nucleic acid), and are separated in a polyacrylamide/urea gel to generate a sequence "ladder". This method is more commonly used than "Maxam-Gilbert" (chemical) sequencing.
DIRECT REPEATS	Identical or related sequences present in two or more copies in the same orientation in the same molecule of DNA; they are not necessarily adjacent.
DMSO	Di methyl suphoxide.
DNA LIGASE	An enzyme (usually from the T4 bacteriophage) that catalyzes formation of a phosphodiester bond between two adjacent bases from double-stranded DNA fragments. RNA ligases also exist, but are rarely used in molecular biology.
DNA POLYMERSASE	An enzyme that synthesizes a double stranded DNA molecule using single strand DNA primer as template.
DOT BLOT	DNA or RNA is simply spotted onto nitrocellulose or nylon membranes, denatured and hybridized with a *probe*. Unlike *Southern* or *northern blots*, there is no separation of the target DNA or RNA by electrophoresis (size), and thus potentially much less specificity.
DOP PCR	Degenerate Oligonucleotide Primer-PCR.

DOWNSTREAM	Identifies sequences proceeding farther in the direction of *expression*; for example, the coding region is downstream from the *initiation codon*, toward the 3' end of an mRNA molecule. Sometimes used to refer to a position within a protein sequence, in which case downstream is toward the *carboxyl* end that is synthesized after the *amino* end during translation.
DUPLEX	A nucleic acid molecule in which two strands are *base paired* with each other.
E	
EDTA	Ethylinediaminetatraacetic acid
ELECTROPORATION	A method for introducing foreign nucleic acid into bacterial or eukaryotic cells that uses a brief, high voltage DC charge that renders the cells permeable to the nucleic acid. Also useful for introducing synthetic peptides into eucaryotic cells.
ELISA	Enzyme Linked Immunosorbent Assay. Very widely used group of methods whereby an enzyme attached to reactant in a system is used to generate a colour signal with an appropriate chromogen and substrate combination. Methods include Indirect, sandwich and competitive systems. Main serological tool for diagnosis. Most methods employ microplate methods utilizing antigens or antibodies passively attached to microplate wells. Separation of sequential reactions is easily made by washing steps. Multichannel spectrophotometers make reading colour rapid.
ELONGATION OR EXTENSION	Phase of PCR cycle following annealing of primer during that the *Taq* polymerase manufactures a strand of DNA. The optimum temperature depends on the enzyme used but is usually between 68-72°C.
ELOSA	Enzyme linked oligonucleotide sorbent assay.
END LABELING	The technique of adding a radioactively labeled group to one end (5' or 3' end) of a DNA strand.
ENDONUCLEASE	Cleaves bonds within a nucleic acid chain; they may be specific for RNA or for single-stranded or double-stranded DNA. A restriction enzyme is a type of endonuclease.
ENHANCER	A eukaryotic *transcriptional* control element that is a DNA sequence that acts at some distance to enhance the activity of a specific *promoter* sequence. Unlike promoter sequences, the position and orientation of the enhancer sequence is generally not important to its activity.

ETHIDIUM BROMIDE	Intercalates within the structure of nucleic acids in such a way that they fluoresce under UV light. Ethidium bromide staining is commonly used to visualize RNA or DNA in agarose gels placed on UV light boxes. Proper precautions are required, because the ethidium bromide is highly mutagenic and the UV light damaging to the eyes. Ethidium bromide is also included in cesium chloride gradients during ultracentrifugation, to separate *supercoiled* circular DNA from linear and *relaxed* circular DNA.
EVOLUTIONARY CLOCK	Defined by the rate at which mutations accumulate within a given gene.
EXON	The portion of a gene that is actually translated into protein.
EXONUCLEASE	An enzyme that hydroylses DNA beginning at one end of a strand, releasing nucleotides one at a time (thus, there are 3' or 5' exonucleases).
EXPRESSION	Usually used to refer to the entire process of producing a protein from a gene, that includes *transcription*, *translation, post-translational modification* and possibly transport reactions.
EXPRESSION VECTOR	A plasmid or phage designed for production of a polypeptide from inserted foreign DNA under specific controls. Often an *inducer* is used. The vector always provides a promoter and often the *transcriptional start site*, *ribosomal binding sequence*, and *initiation codon*. In some cases the product is a *fusion protein*.
F	
FOOTPRINTING	A technique for identifying the site on a DNA (or RNA) molecule that is bound by some protein by virtue of the protection afforded *phosphodiester bonds* in this region against attack by *nuclease* or nucleolytic compounds.
FRAMESHIFT MUTATION	A mutation (deletion or insertion, never a simple substitution) of one or more *nucleotides* but never a multiple of 3 nucleotides, that shortens or lengthens a trinucleotide sequence representing a *codon*; the result is a shift from one *reading frame* to another reading frame. The amino acid sequence of the protein downstream of the mutation is completely altered, and may even be much shorter or longer due to a change in the location of the first *termination (stop) codon*:
FUSION PROTEIN	A product of recombinant DNA in which the foreign gene product is juxtaposed ("fused") to either the *carboxyl-terminal* or *amino-terminal* portion of a polypeptide encoded by the vector itself. Use of fusion proteins often facilitates expression of otherwise lethal products and the purification of recombinant proteins.

G	
GDNA.	Genomic DNA. Double helix containing the genetic blueprint of the organism. DNA sugar phosphate 'backbone' molecules are joined by the pairing of bases, thymine with adenine and cytosine with guanine. Hydrogen bonds form between the complementary bases holding the two strands together.
GEL ELECTROPHORESIS	Amplicons of differing lengths can be separated by their migration through an agarose gel whilst subject to an electrical field. The negatively charged DNA molecules move at a rate that is inversely proportional to their size. The DNA can be visualised by staining with a fluorescent dye and viewing under UV light. The size of the amplicons is predicted by the positioning of the primers on the target DNA and a product or 'band' of the correct size is taken as positive result. Confirmation can be provided by the use of specific probes to check the identity of the band.
GEL SHIFT	A method by which the interaction of a nucleic acid (DNA or RNA) with a protein is detected. The mobility of the nucleic acid is monitored in an agarose gel in the presence and absence of the protein: if the protein binds to the nucleic acid, the complex migrates more slowly in the gel (hence "gel shift"). A "supershift" allows determination of the specific protein, by virtue of a second shift in mobility that accompanies binding of a specific antibody to the nucleic acid-protein complex.
GENE	Generally speaking, the *genomic* nucleotide sequence that codes for a particular polypeptide chain, including relevant *transcriptional* control sequences and *introns* (if a eukaryote). However, the term is often loosely used to refer to only the relevant coding sequence.
GENE CONVERSION	The alteration of all or part of a gene by a homologous donor DNA that is itself not altered in the process.
GENOME	The complete set of genetic information defining a particular animal, plant, organism or virus.
GENOMIC LIBRARY	A DNA library that contains DNA fragments hopefully representing each region of the genome of an organism, virus, etc, cloned into individual vector molecules for subsequent selection and amplification. The DNA fragments are usually very small in size compared with the genome. Such libraries are *cDNA libraries* when prepared from RNA viruses.
GENOTYPE	The genetic constitution of an organism; determined by its nucleic acid sequence. As applied to viruses, the term implies a group of evolutionarily related viruses possessing a defined degree of nucleotide sequence relatedness.
GLP	Good laboratory practice.
GLYCOPROTEIN	A *glycosylated* protein.

GLYCOSYLATION	The covalent addition of sugar moities to N or O atoms present in the side chains of certain amino acids of certain proteins, generally occurring within the Golgi apparatus during secretion of a protein. Glycosylation sites are only partially predictable by current computer searches for relevant motifs in protein sequence. Glycosylation may have profound but very unpredictable effects on the folding, stability, and antigenicity of secreted proteins. Glycosylation is a property of eukaryotic cells, and differs among different cell types (i.e., it may be very different in yeast or insect cells used for protein expression, when compared with Chinese hamster ovary (CHO) cells).
H	
HA	Haemagglutination.
HAIRPIN	A helical (duplex) region formed by base pairing between adjacent (inverted) complementary sequences within a single strand of RNA or DNA.
HETERODUPLEX DNA	Generated by base pairing between complementary single strands derived from different parental *duplex* molecules; heteroduplex DNA molecules occur during genetic *recombination* in vivo and during *hydridization* of different but related DNA strands in vitro. Since the sequences of the two strands in a heteroduplex differ, the molecule is not perfectly base-paired; the *melting* temperature of a heteroduplex DNA is dependent upon the number of mismatched base pairs.
HOMOLOGOUS RECOMBINATION	The exchange of sequence between two related but different DNA (or RNA) molecules, with the result that a new "chimeric" molecule is created. Several mechanisms may result in recombination, but an essential requirement is the existence of a region of *homology* in the recombination partners. In DNA recombination, breakage of single strands of DNA in the two recombination partners is followed by joining of strands present in opposing molecules, and may involve specific enzymes. Recombination of RNA molecules may occur by other mechanisms.
HOMOLOGY	Indicates similarity between two different *nucleotide* or *amino acid* sequences, often with potential evolutionary significance. It is probably better to use more quantitative and descriptive terms such as nucleotide "identity" or, in the case of proteins, amino acid "identity" or "relatedness" (the latter refers to the presence of amino acids residues with similar polarity/charge characteristics at the same position within a protein).
HYBRIDISATION	The process of *base pairing* leading to formation of *duplex* RNA or DNA or RNA-DNA molecules.

HOT START	To minimise mismatched primers interfering with the PCR the reagents are heated to above the annealing temperature before all the reaction components are allowed to mix. This is often achieved by using wax to separate the primers from the enzyme, the reaction components can mix only when the wax melts thus avoiding the elongation of non-specifically bound primers. Alternatively, enzymes are now available that will only elongate at elevated temperatures.
HYDROPHILICITY PLOT	A computer plot that examines the relative summed hyrophobicity/hydrophilicity of adjacent *amino acid* sidechains (usually within a moving window of about 6 amino acid residues) along the primary sequence of a polypeptide chain. Values for the contribution of sidechains of each the 20 common amino acids to hydrophobicity/hydrophilicity have been developed by Hopp & Woods, and Kyte & Doolittle, and these plots are often named after these workers. Generally, hydrophobic regions of proteins are considered likely to be in the interior of the native protein, while hydrophilic domains are likely to be exposed on the surface and thus possibly antigenic sites (epitopes). At best, these are crude predictions.
I	
IMMUNOBLOT	See WESTERN BLOT.
INDUCER	A small molecule, such as IPTG, that triggers gene transcription by binding to a regulator protein, such as LacZ.
INHIBITION.	The exponential nature of PCR means that a small change in the efficiency of each cycle can result in a large difference to the final product. Inhibition may result from chemical interference in the action of polymerase or other components of the reaction and could be removed by further purification of the sample. Alternatively, a large excess of DNA in the sample may prevent efficient amplification of the target and further dilution of sample is required. Checks for inhibition can be made by adding a known target to the sample to demonstrate the efficiency of the amplification.
INITIATION CODON	The *codon* at which translation of a polypeptide chain is initiated. This is usually the first AUG triplet in the *mRNA* molecule from the 5' end, where the ribosome binds to the cap and begins to scan in a 3' direction. However, the surrounding sequence context is important and may lead to the first AUG being bypassed by the scanning ribosome in favor of an alternative, downstream AUG. Also called a "start codon". Occasionally other codons may serve as initiation codons, e.g. UUG.
INSERT	Foreign DNA placed within a vector molecule.

INSERTION SEQUENCE	A small bacterial transposon carrying only the genetic functions involved in *transposition*. There are usually *inverted repeats* at the ends of the insertion sequence.
INTRON	Intervening sequences in eukaryotic genes that do not encode protein but which are *transcribed* into RNA. Removed from *pre-mRNA* during nuclear *splicing* reactions.
INVERTED REPEATS	Two copies of the same or related sequence of DNA repeated in opposite orientation on the same molecule (contrast with DIRECT REPEATS). Adjacent inverted repeats constitute a *palindrome*.
IN SITU PCR.	PCR of DNA within histological specimens.
IRE PCR	Interspersed repetitive element (IRE)-PCR
ISO	A network of national standards institutes from 148 countries working in partnership with international organizations, governments, industry, business and consumer representatives.
ISOTHERMAL AMPLIFICATION.	Amplification step proceeds at one temperature and does not require thermocycling apparatus.
K	
KILOBASE	Unit of 1000 nucleotide bases, either RNA or DNA.
KLENOW FRAGMENT	The large fragment of E. coli DNA polymerase I that lacks 5' -> 3' exonuclease activity. Very useful for sequencing reactions, that proceed in a 5' > 3' fashion (addition of nucleotides to templated free 3' ends of primers).
KNOCK-OUT	The excision or inactivation of a gene within an intact organism or even animal (e.g., "knock-out mice"), usually carried out by a method involving *homologous recombination*.
L	
LCR	Ligase chain reaction. Amplification of DNA by repeated ‘cycling’ reactions using ligase enzyme to ‘ligate’ or ‘join’ pairs of primers. Covered by patents owned by Abbott Laboratories (USA). A range of diagnostic kits is available.
LIBRARY	A set of cloned fragments together representing with some degree of redundancy the entire genetic complement of an organism (see cDNA LIBRARY, GENOMIC LIBRARY).
LIGASE	See DNA LIGASE.
LIGATION	See DNA LIGASE.
LINKAGE	The tendency of genes to be inherited together as a result of their relatively close proximity on the same chromosome, or location on the same plasmid.

LINKER	A short *oligodeoxyribonucleotide*, usually representing a specific *restriction endonuclease recognition sequence*, that may be *ligated* onto the termini of a DNA molecule to facilitate cloning. Following the ligation reaction, the product is digested with the endonuclease, generating a DNA fragment with the desired *sticky* or *blunt ends*.
LIPOFECTIN	A commercially marketed liposome suspension that is mixed with DNA or RNA to facilitate uptake of the nucleic acid by eukaryotic cells (see TRANSFECTION).
LONG PCR	PCR in which the product is several kilobases long. Using cocktails of enzymes products of up to 40 kb have been accurately manufactured.
LSD	
M	
mRNA	Messenger RNA.
MELTING	The dissociation of a duplex nucleic acid molecule into single strands, usually by increasing temperature. See DENATURATION.
$MgCl_2$	Magnesium chloride.
Mg^{2+}	Magnesium ion.
MISSENSE MUTATION	A nucleotide mutation that results in a change in the amino acid sequence of the encoded protein (contrast with SILENT MUTATION).
M-MuLV	< Moloney murine leukemia virus RT.
MOTIF	A recurring pattern of short sequence of DNA, RNA, or protein, that usually serves as a recognition site or active site. The same motif can be found in a variety of types of organisms. MRNA – A cytoplasmic RNA that serves directly as the source of code for protein synthesis.
MULTICISTRONIC MESSAGE	An *mRNA transcript* with more than one *cistron* and thus encoding more than one *polypeptide*. These generally do not occur in eukaryotic organisms, due to differences in the mechanism of translation initiation.
MULTICOPY PLASMIDS	Present in bacteria at amounts greater than one per chromosome. Vectors for cloning DNA are usually multicopy; there are sometimes advantages in using a single copy plasmid.
MULTIPLE CLONING SITE	An artificially constructed region within a *vector* molecule that contains a number of closely spaced *recognition sequences* for *restriction endonucleases*. This serves as a convenient site into which foreign DNA may be inserted.

MULTIPLEX PCR	PCR with two or more independent sets of primers in the same reaction tube.
N	
N TERMINUS	See AMINO TERMINUS.
NATIVE GEL	An electrophoresis gel run under conditions that do not denature proteins (i.e., in the absence of SDS, urea, 2-mercaptoethanol, etc.).
NESTED PCR (n-PCR)	A very sensitive method for amplfication of DNA, that takes part of the product of a single *PCR* reaction (after 30-35 cycles), and subjects it to a new round of PCR using a different set of *PCR primers* which are nested within the region flanked by the original primer pair (see POLYMERASE CHAIN REACTION).
NICK	In *duplex* DNA, this refers to the absence of a *phosphodiester bond* between two adjacent *nucleotides* on one strand.
NICK TRANSLATION	A method for introducing labeled *nucleotides* into a double-stranded DNA molecule that involves making small *nicks* in one strand with DNase, and then repairing with DNA polymerase I.
NONCONSERVATIVE SUBSTITUTION	A mutation that results in the substitution of one *amino acid* within a polypeptide chain with an amino acid belonging to a different polarity/charge group.
NONSENCE CODON	See STOP CODON.
NONSENSE MUTATION	A change in the sequence of a nucleic acid that causes a *nonsense* (*stop* or *termination*) *codon* to replace a codon representing an amino acid.
NONTRANSLATED RNA (NTR)	The segments located at the 5' and 3' ends of a mRNA molecule that do not encode any part of the polyprotein; may contain important translational control elements.
NORTHERN BLOT	RNA molecules are separated by electrophoresis (usually in an agarose gel) on the basis of size, then transferred to a solid-phase support (nitrocellulose paper or suitable other membrane) and detected by *hybridisation* with a labeled probe (see SOUTHERN BLOT, WESTERN BLOT).
NUCLEOSIDE	The composite sugar and *purine or pyrimidine base* that is present in *nucleotides* which are the basic building blocks of DNA and RNA. Compare with NUCLEOTIDE: Nucleoside = Base + Sugar.
NUCLEOTIDE	The composite phosphate, sugar, and *purine* or *pyrimidine base* that are the basic building blocks of the nucleic acids DNA and RNA. The five nucleotides are adenylic acid, guanylic acid (contain *purine bases*), and cytidylic acid, thymidylic acid, and uridylic acid (contain *pyrimidine bases*).

NASBA	Nucleic Acid Sequence-Based Amplification. Amplification of RNA using RNA polymerase. Developed by Cangene Corporation (Canada), also licensed to Organon Teknika (Belgium). Nucleotide = Base + Sugar + Phosphate (1, 2, or 3).
O	
OIE	Office International des Epizootie, World Organisation for Animal health. 12, rue de Prony, 75017 Paris, France. Email : oie@oie.int; Web Site: www.oie.int
OLIGODEOXYRIBONUCLEOTIDE	A short, single-stranded DNA molecule, generally 15-50 *nucleotides* in length, that may be used as a *primer* or a *hybridisation probe*. Oligodeoxyribonucleotides are synthesized chemically under automated conditions.
OLIGONUCLEOTIDE -	See OLIGODEOXYRIBONUCLEOTIDE
ONCOGENE	One of a number of genes believed to be associated with the malignant transformation of cells; originally identified in certain oncogenic retroviruses (*v-onc*) but also present in cells (*c-onc*).
ONE-TUBE NESTED PCR	Both reactions in a nested PCR are performed concurrently in one tube without the need to add further reagents.
OPEN READING FRAME	A region within a reading frame of an mRNA molecule that potentially encodes a *polypeptide*; and that does not contain a *translational stop codon* (see READING FRAME).
OPERATOR	The site on DNA at which a *repressor* protein binds to prevent *transcription* from initiating at the adjacent *promoter*.
OPERON	A complete unit of bacterial gene expression and regulation, including the structural gene or genes, regulator gene(s), and control elements in DNA recognized by regulator gene products(s).
ORIGIN	A site within a DNA sequence of a chromosome, plasmid, or non-integrated virus at which replication of the DNA is initiated.
OVERHANG	A terminus of a *duplex* DNA molecule that has one or more unpaired nucleotides in one of the two strands (hence either a 3' or 5' overhang). Cleavage of DNA with many restriction endonucleases leaves such overhangs (see STICKY END).
P	
PACKAGE	In recombinant DNA procedures, refers to the step of incorporation of *cosmid* or other lambda *vector* DNA with an *insert* into a *phage* head for transduction of DNA into host.
PAGE	Polyacrylamide gel electrophoresis.

PALINDROMIC SEQUENCE	A nucleotide sequence that is the same when read in either direction, usually consisting of adjacent inverted repeats. *Restriction endonuclease recognition sites* are palindromes:
PCR	See POLYMERASE CHAIN REACTION.
PCR-ELISA	PCR Enzyme linked immunosorbent assay.
pfu	Plaque forming units.
PHENOTYPE	The appearance of other characteristics of an organism resulting from the interaction of its genetic constitution with the environment.
PHOSPHATASE, ALKALINE	An enzyme that catalyzes the hydrolysis of phosphomonoesters of the 5' nucleotides. Used to dephosphorylate (remove phosphate groups from) the 5' ends of DNA or RNA molecules, to facilitate 5' end-labeling with 32P added back by T4 polynucleotide kinase; or to dephosphorylate the 5' ends of DNA molecules to prevent unwanted *ligation* reactions during cloning.
PHOSPHODIESTER BOND	The covalent bond between the 3' hydroxyl in the sugar ring of one nucleotide and the 5' phosphate group of the sugar ring of the adjacent nucleotide residue within a nucleic acid: 5'-Ribose- 3' - O - P(O)2 - O - 5' -Ribose - 3' - etc.
PHOSPHORYLATION	The addition of a phosphate monoester to a macromolecule, catalyzed by a specific kinase enzyme. With respect to proteins, certain *amino acid* side chains (serine, threonine, tyrosine) are subject to phosphorylation catalyzed by protein kinases; altering the phosphorylation status of a protein may have dramatic effects on its biologic properties, and is a common cellular control mechanism. With respect to DNA, 5' ends must be phosphorylated for *ligation*.
PLASMID	An extrachromosomal, usually circular, double-stranded DNA that is capable of *replication* within a cell, and which usually contains and expresses genes encoding resistance to antibiotics. By strict definition, a plasmid is not essential to the life of the cell.
POINT MUTATION	A single nucleotide substitution within a gene; there may be several point mutations within a single gene. Point mutations do not lead to a shift in reading frames, thus at most cause only a single amino acid substitution (see FRAMESHIFT MUTATION).
POLYACRYLAMIDE GEL (PAGE)	Used to separate proteins and smaller DNA fragments and *oligonucleotides* by electrophoresis. When run under conditions that denature proteins (i.e., in the presence of 2-mercaptoethanol, SDS, and possibly urea), molecules are separated primarily on the basis of size.

POLY-A TRACK	A lengthy adenylic acid polymer (RNA) that is covalently linked to the 3' end of newly synthesized *mRNA* molecules in the nucleus.
POLYMERASE CHAIN REACTION (PCR)	A DNA amplification reaction involving multiple (30 or more) cycles of *primer annealing, extension*, and denaturation, usually using a heat-stable DNA *polymerase* such as *Taq polymerase*. Paired primers are used, that are complementary to opposing strands of the DNA and which flank the area to be amplified. Under optimal conditions, single DNA sequence can be amplified a million-fold. The basic PCR process is covered by patents owned by Hoffman-La Roche Inc and Hoffman-La Roche Ltd. A range of diagnostic kits is available.
POLYMERASE	An enzyme that catalyzes the addition of a *nucleotide* to a nucleic acid molecule. There is a wide variety of RNA and DNA polymerases that have a wide range of specific activities and that operate optimally under different conditions. In general, all polymerases require *templates* upon which to build a new strand of DNA or RNA; however, DNA polymerases also require a *primer* to initiate the new strand, while RNA polymerases start synthesis at a specific *promoter* sequence.
POLYMORPHISM	Variation within a DNA or RNA sequence.
POLYNUCLEOTIDE KINASE	Enzyme which catalyzes the transfer of the terminal phosphate of ATP to 5' hydroxyl termini of polynucleotides, either DNA or RNA. Usually derived from T4 bacteriophage. pre-mRNA - An RNA molecule that is *transcribed* from chromosomal DNA in the nucleus of eukaryotic cells, and subsequently processed through *splicing* reactions to generate the mRNA that directs protein synthesis in the cytoplasm.
POST-TRANSLATIONAL MODIFICATION	Modifications made to a polypeptide molecule after its initial synthesis, this includes proteolytic cleavages, *phosphorylation, glycosylation*, carboxylation, addition of fatty acid moieties, etc.
PRE-MRNA	An RNA molecule that is *transcribed* from chromosomal DNA in the nucleus of eukaryotic cells, and subsequently processed through *splicing* reactions to generate the mRNA that directs protein synthesis in the cytoplasm.
PRIMARY STRUCTURE	Refers to the sequence of *amino acid residues* or *nucleotides* within protein or nucleic acid molecules, respectively (also see SECONDARY and TERTIARY STRUCTURE).

PRIMER	An *oligonucleotide* that is *complementary* to a specific region within a DNA or RNA molecule, and that is used to prime (initiate) synthesis of a new strand of complementary DNA at that specific site, in a reaction or series of reactions catalyzed by a *DNA polymerase*. The newly synthesized DNA strand will contain the primer at its 5' end. Typically, primers are chemically synthesized oligonucleotides 15-50 nucleotides in length, selected on the basis of a known sequence (also known as 15-50mers primers). However, "random primers" (shorter oligonucleotides, about 6 nucleotides in length, and comprising all possible sequences) may be used to prime DNA synthesis from DNA or RNA of unknown sequence. but probably serves to enhance stability of the RNA. Is frequently used to select mRNA for cloning purposes by *annealing* to a column containing a matrix bound to poly-uridylic acid.
PRIMER-DIMER	At room temperature temporary association of primers may take place allowing the *Taq* enzyme to polymerise small strands of DNA called primer-dimers. Can give a false positive result if the detection of amplicons is via labelled primers.
PRIMER EXTENSION	A reaction in which DNA is *reverse transcribed* from an RNA *template* to which a specific oligonucleotide *primer* has been *annealed*. The new cDNA product is an extension of the primer, that is synthesized at the 3' end of the primer in a direction extending toward the 5' end of the RNA. This reaction is useful for exploring the extreme 5' end of RNA molecules.
PROBE	Usually refers to a DNA or RNA molecule that has been labeled with 32P or with *biotin*, to facilitate its detection after it has specifically *hybridized* with a target DNA or RNA sequence. However, the term may also refer to antibody probes used in *western* blots.
PROCESSING	With respect to proteins, generally used to refer to proteolytic *post-translational modifications* of a polypeptide. In the case of RNA, processing may involve the addition of a 5' *cap* and 3' *poly-A* tracks as well as *splicing* reactions in the nucleus.
PROCESSIVITY	The extent to which an RNA or DNA *polymerase* adheres to a *template* before dissociating, determines the average length (in kilobases) of the newly synthesized nucleic acid strands. Also applies to the action of *exonucleases* in digesting from the ends to the middle of a nucleic acid.

PROMOTER	A specific sequence within a double-stranded DNA molecule that is recognized by an RNA *polymerase*, that binds to it and uses it to begin transcribing the DNA *template* into a new RNA. The location and orientation of the *promoter* within a DNA molecule determines the start site of the new RNA. Other proteins (e.g. transcriptional activators such as *sigma factor*) are usually required for an RNA polymerase to recognize a promoter (see TRANSCRIPTION).
PROTO-ONCOGENE	A cellular *oncogene*-like sequence that is thought to play a role in controlling normal cellular growth and differentiation.
PSEUDOGENE	Inactive but stable components of the genome that derived by duplication and mutation of an ancestral, active gene. Pseudogenes can serve as the donor sequence in *gene conversion* events.
PSEUDOREVERTANT	A mutant virus or organism that has recovered a wildtype phenotype due to a second-site mutation (potentially located in a different region of the genome, or involving a different polypeptide) that has eliminated the effect of the initial mutation.
PSEUDOKNOT	A feature of RNA *tertiary structure*; best visualized as two overlapping *stem-loops* in that the loop of the first stem-loop participates as half of the stem in the second stem-loop.
PURINE BASES	Adenine (A) or Guanine (G) (see NUCLEOTIDE).
PULSED-FIELD GEL ELECTROPHORESIS (PFGE)	Separation of large (>50 kb) pieces of DNA, including complete chromosomes and genomes, by rapidly alternating the direction of electrophoretic migration in agarose gels.
PYRIMIDINE BASES	Cytosine (C), Thymine (T) or Uracil (U) (see NUCLEOTIDE).
Q	
Q-BETA REPLICASE	An isothermal nucleic acid amplification system that uses the enzyme Q-beta replicase to replicate an RNA probe. An automated diagnostic system is being developed by Vysis Inc. (USA).

R	
RACE	Rapid Amplification of Complementary Ends
READING FRAME	Refers to a polypeptide sequence potentially encoded by a single-stranded *mRNA*. Because *codons* are nucleotide triplets, each mRNA has 3 reading frames (each nucleotide can participate in 3 codons, at the 1st, 2nd, and 3rd base position). Duplex DNA strands have 6 reading frames, 3 in each strand (see OPEN READING FRAME): AlaSerProLeuVal . 1st reading frame ProAlaProTERTrp 2nd reading frame: TER = Stop GlnProProSerGly 3rd reading frame.
RECOGNITION SEQUENCE	A specific *palindromic sequence* within a double-stranded DNA molecule that is recognized by a *restriction endonuclease*, and at which the restriction endonuclease specifically cleaves the DNA molecule.
RECOMBINATION-REPAIR	A mode of filling a gap in one strand of duplex DNA by retrieving a homologous single strand from another duplex. Usually the underlying mechanism behind *homologous recombination* and *gene conversion*.
RELAXED DNA	See SUPERCOIL.
REPLICATION	The copying of a nucleic acid molecule into a new nucleic acid molecule of similar type (i.e., DNA > DNA, or RNA > RNA
REPORTER GENE	The use of a functional enzyme, such as beta-galactosidase, luciferase, or chloramphenicol acetyltransderase, downstream of a gene, promoter, or translational control element of interest, to more easily identify successful introduction of the gene into a host and to measure transcription and/or translation.
REPRESSION	Inhibition of transcription (or translation) by the binding of a repressor protein to a specific site on DNA (or *mRNA*
RESIDUE	As applied to proteins, what remains of an *amino acid* after its incorporation into a peptide chain, with subsequent loss of a water molecule (see PEPTIDE BOND).
RESTRICTION ENDONUCLEASE (RE)	A bacterial enzyme that recognizes a specific *palindromic sequence (recognition sequence)* within a double-stranded DNA molecule and then catalyzes the cleavage of both strands at that site. Also called a restriction enzyme. Restriction endonucleases may generate either *blunt* or *sticky ends* at the site of cleavage.

RESTRICTION FRAGMENT LENGTH POLYMORPHISM (RFLP)	Variations in the lengths of fragments of DNA generated by digestion of different DNAs with a specific *restriction endonuclease*, reflecting genetic variation (*polymorphism*) in the DNAs.
RESTRICTION FRAGMENTS	DNA fragments generated by digestion of a DNA preparation with one or more *restriction endonucleases*; usually separated by *agarose gel electrophoresis* and visualized by *ethidium bromide* staining under UV light (or alternatively subjected to *Southern blot analysis*).
RESTRICTION MAP	A linear array of sites on a particular DNA that are cleaved by various selected *restriction endonucleases.*
RETICULOCYTE LYSATE	A lysate of rabbit reticulocytes, that has been extensively digested with micrococcal nuclease to destroy the reticulocyte *mRNAs*. With the addition of an exogenous, usually synthetic, mRNA, *amino acids* and a source of energy (ATP), the translational machinery of the reticulocyte (*ribosomes*, eukaryotic translation factors, etc.) will permit in vitro *translation* of the added mRNA with production of a new *polypeptide*. This is only one of several available in vitro translation systems.
REVERSE TRANSCRIPTASE (RT)	A *DNA polymerase* that copies an RNA molecule into single-stranded cDNA; usually purified from retroviruses.
REVERSE TRANSCRIPTION	Copying of an RNA molecule into a DNA molecule.
RT-PCR	Real time PCR
RIBONUCLEASE (RNASE)	An enzyme that catalyzes the hydrolysis of RNA. There are many different RNases, some of the more important include
RNASE H	Degrades the RNA part of RNA:DNA hybrids.
RIBOSOMAL BINDING SEQUENCE (SHINE-DALGARNO SEQUENCE)	In prokaryotic organisms, part or all of the polypurine sequence AGGAGG located on *mRNA* just upstream of an AUG *initiation codon*; it is complementary to the sequence at the 3' end of 16S rRNA; and involved in binding of the ribosome to *mRNA*. The *internal ribosomal entry site* found in some viruses may be an analogous eukaryotic genetic element.
RIBOSOME	A complex ribonucleoprotein particle (eukaryotic ribosomes contain 4 RNAs and at least 82 proteins) which is the "machine" which translates *mRNA* into protein molecules. In eukaryotic cells, ribosomes are often in close proximity to the endoplasmic reticulum.
RIBOZYME	A catalytically active RNA. A good example is the hepatitis delta virus RNA that is capable of self-cleavage and self-ligation in the absence of protein enzymes.
RNA POLYMERASE	A polymerase that synthesizes RNA (see POLYMERASE).

RNA SPLICING	A complex and incompletely understood series of reactions occurring in the nucleus of eukaryotic cells in which *pre-mRNA transcribed* from chromosomal DNA is processed such that noncoding regions of the pre-mRNA (*introns*) are excised, and coding regions (*exons*) are covalently linked to produce an *mRNA* molecule ready for transport to the cytoplasm. Because of splicing, eukaryotic DNA representing a gene encoding any given protein is usually much larger than the mRNA from which the protein is actually *translated.*
RNASE -	see RIBONUCLEASE
rRNA	Ribosomal RNA (four sizes in humans: 5S, 5.8S, 18S, and 28S); RNA component of the *ribosome* that may play catalytic roles in *translation.*
RT	Reverse Transcriptase.
RT/PCR REACTION	A series of reactions that result in RNA being copied into DNA and then amplified. A single *primer* is used to make single-stranded *cDNA* copies from an RNA *template* under direction of *reverse transcriptase.* A second primer *complementary* to this "first strand" cDNA is added to the reaction mix along with *Taq polymerase*, resulting in synthesis of double-stranded DNA. The reaction mix is then cycled (denaturation, *annealing* of primers, *extension*) to amplify the DNA by conventional PCR.
RUNOFF TRANSCRIPT	RNA that has been synthesized from plasmid DNA (usually by a RNase A Cleaves ssRNA 3' of pyrimidines RNase T1 Cleaves ssRNA at guanasine nucleotides RNase V1 Cleaves dsRNA (helical regions)bacteriophage *RNA polymerase* such as *T7* or *SP6*) and which terminates at a specific 3' site because of prior cleavage of the plasmid DNA with a *restriction endonuclease.* **RNA**. Ribonucleic acid. Single strand nucleic acid composed of a phosphate ribose 'backbone' and the nucleotides: adenine, guanine, cytosine and uracil.
MRNA	Messenger RNA, carries genetic information from DNA to the ribosome where it is translated into the corresponding protein.
S	
SEMI OR HEMI-NESTED PCR	If a full nested PCR (using two internal primers) cannot be performed, sensitivity and specificity can be improved by using just one 'inner' primer in conjunction with one of the 'outer' primers from the first reaction.
S1 NUCLEASE	An enzyme that digests single-stranded DNA or RNA.
SDS-PAGE	Denaturing protein gel electrophoresis (see POLYACRYLAMIDE GEL ELECTROPHORESIS).

SELECTION	The use of particular conditions, such as the presence of ampicillin, to allow survival only of cells with a particular *phenotype*, such as production of beta-lactamase.
SEQUENCE POLYMORPHISM	See POLYMORPHISM.
SHOTGUN CLONING OR SEQUENCING	Cloning of an entire genome or large piece of DNA in the form of randomly generated small fragments. The individual sequences obtained from the clones will be used to construct *contigs*.
SHUTTLE VECTOR	A small *plasmid* capable of *transfection* into both prokaryotic and eukaryotic cells.
SIGMA FACTOR	Certain small ancillary proteins in bacteria that increase the binding affinity of RNA polymerase to a promoter. Different sigma factors recognize different promoter sequences.
SIGNAL PEPTIDASE	An enzyme present within the lumen of the endoplasmic reticulum that proteolytically cleaves a secreted protein at the site of a *signal sequence.*
SIGNAL SEQUENCE	A hydrophobic amino acid sequence that directs a growing peptide chain to be secreted into the endoplasmic reticulum.
SILENT MUTATION	A nucleotide substitution (never a single deletion or insertion) that does not alter the amino acid sequence of an encoded protein due to the *degeneracy* of the genetic code. Such mutations usually involve the third base (*wobble* position) of *codons.*
SITE-DIRECTED MUTAGENESIS	The introduction of a mutation, usually a *point mutation* or an insertion, into a particular location in a cloned DNA fragment. This mutated fragment may be used to "*knock out*" a gene in the organism of interest by *homologous recombination.*
SITE-SPECIFIC RECOMBINATION	Occurs between two specific but not necessarily homologous sequences. Usually catalyzed by enzymes not involved in general or *homologous recombination.*
SOP	Standard Operating procedure
SOUTHERN BLOT	DNA is separated by electrophoresis (usually in *agarose gels*), then transferred to nitrocellulose paper or other suitable solid-phase matrix (e.g., nylon membrane), and denatured into single strands so that it can be *hybridized* with a specific *probe.* The Southern blot was developed by E.M. Southern, a molecular biologist in Edinburgh. *Northern* and *western* blots were given contrasting names to reflect the different target substances (RNA and proteins, respectively) that are subjected in these procedures to electrophoresis, blotting and subsequent detection with specific probes.
SOUTHWESTERN BLOT	The binding of protein to a nucleic acid on a matrix similar to what is done for western, northern, and southern blots. This technique is used to identify DNA binding proteins and the recognition sites for these proteins.

SP6 RNA POLYMERASE	A bacteriophage *RNA polymerase* that is commonly used to transcribe *plasmid* DNA into RNA. The plasmid must contain an SP6 *promoter* upstream of the relevant sequence.
SPLICING	see RNA SPLICING.
ss	Single stranded.
ss DNA	Single stranded DNA.
ss RNA	Single stranded RNA.
START CODON	See INITIATION CODON.
STEM-LOOP	A feature of RNA *secondary structure*, in which two complementary, inverted sequences that are separated by a short-intervening sequence within a single strand of RNA base pair to form a "'stem" with a "loop" at one end. Similar to a *hairpin*, but these usually have very small loops and longer stems.
STICKY END	Terminus of DNA molecule that has either a 3' or 5' overhang, and which typically results from a cut by a *restriction endonuclease*. Such termini are capable of specific ligation reactions with other termini that with complementary overhangs. A sticky end can be "blunt ended" by the removal of an overhang, or a "filling in" reaction that adds additional nucleotides complementary to overhang (see BLUNT END).
SDA	Strand Displacement Assay. An isothermal nucleic acid amplification system where cDNA product is synthesised from an RNA target. Covered by patents owned by Becton Dickinson (USA).
T	
Ta	Annealing temperature.
TAE	Tris acetic acid EDTA.
TARGET	The specific piece of DNA or RNA to be amplified by the PCR. The sensitivity of an assay can be enhanced if multiple copies of the target are present. e.g. a repeated sequence or mRNA molecules.
TAQ	Thermostable DNA polymerase, originally isolated from *Thermus aquaticus,* an organism that resides in hot springs.
TAQMAN.	5' nuclease fluorogenic assay. For simple and rapid detection of PCR products. An oligonucleotide probe that is specific for the target to be amplified is labelled with a fluorescent tag and a quenching molecule. During the extension step of a PCR the Taq enzyme will disrupt probe bound to the target separating the fluorescent tag from its quencher molecule thus permitting fluorescence.

***Taq* POLYMERASE**	A *DNA polymerase* that is very stable at high temperatures, isolated from the thermophilic bacterium *Thermus aquaticus*. Very useful in *PCR* reactions which must cycle repetitively through high temperatures during the denaturation step.
TBE	Tris boric acid EDTA.
TEMPLATE	A nucleic acid strand, upon which a *primer* has *annealed* and a nascent RNA strand is being extended.
TERMINATION CODON	See STOP CODON
TERMINATOR	A sequence *downstream* from the 3' end of an *open reading frame* that serves to halt *transcription* by the RNA polymerase. In bacteria these are commonly sequences that are *palindromic* and thus capable of forming *hairpins*. Sometimes termination requires the action of a protein, such as Rho factor in *E. coli.*
THERMAL CYCLER	Although PCR may be performed using manual transfer between a series of water or oil baths, for reproducible results or diagnostic work a reliable programmable thermal cycling machine is essential. There are now many machines to choose from with a wide range of specifications and costs. A patent on the use of thermal cycling machines for PCR is held by the Perkin Elmer Corporation (USA). It should be remembered that heating and cooling parameters vary with the design of the instrument and a PCR that has been optimised for one machine may not function as efficiently on a different type of thermal cycler.
THREE ROOM STRATEGY.	In order to minimise the risks of contamination 3 workstations should be used, each in a separate room. A 'Clean' room for preparation of PCR reagents, a 'Grey' area for preparation of samples and positive controls and a 'Dirty' room for post-amplification work and detection of amplicons.
Tm	The midpoint of the temperature range over which DNA is melted or denatured by heat; the temperature at that a *duplex* nucleic acid molecule is 50% *melted* into single strands, it is dependent upon the number and proportion of G-C *base pairs* as well as the ionic conditions. Often referred to as a measure of the thermal stability of a nucleic acid *probe*:target sequence hybrid. To calculate the approximate Tm for an oligonucleotide count the numbers of each nucleotide. $Tm = 2(A+T) + 4\ (G + C)$
TMA	Transcription-mediated amplification, an isothermal nucleic acid amplification system. Amplification of RNA using RNA polymerase. Covered by patents owned by Gen-Probe Incorporated (USA). A range of diagnostic kits is available.

T7 RNA POLYMERASE	A bacteriophage *RNA polymerase* that is commonly used to transcribe plasmid DNA into RNA. The plasmid must contain a T7 *promoter* upstream of the relevant sequence.
TRANS	As used in molecular biology, an interaction that involves two sites that are located on separate molecules.
TRANSCRIPT	A newly made RNA molecule that has been copied from DNA.
TRANSCRIPTION	The copying of a DNA template into a single-stranded RNA molecule. The processes whereby the transcriptional activity of eukaryotic genes are regulated are complex, involve a variety of accessory transcriptional factors that interact with *promoters* and *polymerases*, and constitute one of the most important areas of biological research today.
TRANSCRIPTION/TRANSLATION REACTION	An in vitro series of reactions, involving the synthesis (*transcription*) of an mRNA from a *plasmid* (usually with *T7* or *SP6 RNA polymerase*), followed by use of the mRNA to program *translation* in a cell-free system such as a rabbit *reticulocyte lysate*. The *polypeptide* product of translation in usually labelled with [35S]-methionine, and examined in an *SDS-PAGE* gel with or without prior *immunoprecipitation*. This series of reactions permits the synthesis of a polypeptide from DNA in vitro.
TRANSCRIPTIONAL START SITE	The nucleotide of a gene or cistron at which *transcription* (RNA synthesis) starts; the most common triplet at which transcription begins in *E. coli* is CAT. *Primer extension* identifies the transcriptional start site.
TRANSFECTION	The process of introducing foreign DNA (or RNA) into a host organism, usually a eukaryotic cell.
TRANSFORMATION	Multiple meanings. With respect to cloning of DNA, refers to the transformation of bacteria (usually to specific antibiotic resistance) due to the uptake of foreign DNA. With respect to eukaryotic cells, usually means conversion to less-restrained or unrestrained growth.
TRANSGENE	A foreign gene that has been introduced into the germ line of an animal species.
TRANSGENIC	An animal (usually a mouse) or plant into which a foreign gene has been introduced in the germ line. An example: transgenic mice expressing the human receptor for poliovirus are susceptible to human polioviruses.

TRANSITION	A nucleotide substitution in which one pyrimidine is replaced by the other pyrimidine, or one purine replaced by the other purine (e.g., A is changed to G, or C is changed to T) (contrast with TRANSVERSION).
TRANSLATION	The process whereby *mRNA* directs the synthesis of a protein molecule; carried out by the *ribosome* in association with a host of translation initiation, elongation and termination factors. Eukaryotic genes may be regulated at the level of translation, as well as the level of *transcription.*
TRANSLOCATION	The process by which a newly synthesized protein is directed toward a specific cellular compartment (i.e, the nucleus, the endoplasmic reticulum).
TRANSPOSON	A transposable genetic element; certain sequence elements that are capable of moving from one site to another in a DNA molecule without any requirement for sequence relatedness at the donor and acceptor sites. Many transposons carry antibiotic resistance determinants and have *insertion sequences* at both ends, and thus have two sets of *inverted repeats*.
TRANSPOSITION	The movement of DNA from one location to another location on the same molecule, or a different molecule within a cell.
TRANSVERSION	A nucleotide substitution in which a purine replaces a pyrimidine, or vice versa (e.g., A is changed to T, or T is changed to G) (see TRANSITION).
TRIPLET	A three-nucleotide sequence; a *codon.*
tRNA	Small, tightly folded RNA molecules which act to bring specific amino acids into *translationally* active *ribosomes* in a fashion that is dependent upon the *mRNA* sequence. One end of the tRNA molecule recognizes the nucleotide triplet which is the *codon* of the mRNA, while the other end (when activated) is covalently linked to the relevant *amino acid.*
U	
UNG	Uracil N-glycosylase.
UNTRANSLATED RNA	See NONTRANSLATED RNA.
UPSTREAM	Identifies sequences located in a direction opposite to that of expression; for example, the bacterial *promoter* is upstream of the *initiation codon*. In an mRNA molecule, upstream means toward the 5' end of the molecule. Occasionally used to refer to a region of a polypeptide chain that is located toward the amino terminus of the molecule.
UV	Ultra violet.

V	
VECTOR	A *plasmid, cosmid*, bacteriophage, or virus that carries foreign nucleic acid into a host organism.
W	
WESTERN BLOT	Proteins are separated by *SDS-PAGE*, then electrophoretically transferred to a solid-phase matrix such as nitrocellulose, then probed with a labelled antibody (or a series of antibodies).
WHO	World Health Organisation.
WILDTYPE	The native or predominant genetic constitution before mutations, usually referring to the genetic constitution normally existing in nature.
WOBBLE POSITION	The third base position within a *codon*, which can often (but not always) be altered to another nucleotide without changing the encoded amino acid (see DEGENERACY).
5' AND 3'.	The orientation of nucleic acids is indicated by the terms 3' or 5'. These refer to structures within the phosphate/sugar backbone of the molecules. The 5' end has a phosphate group whereas the hydroxyl (OH) end of the molecule is designated 3'. Single strand DNA sequences are normally written from the 5' end of the molecule to 3' end. In double-stranded DNA the lower, second strand will lie in the opposite direction from 3' to 5'. *Taq* enzyme manufactures DNA from the 3' end of the primer.

INDEX

Index

Index

GPSR Compliance

The European Union's (EU) General Product Safety Regulation (GPSR) is a set of rules that requires consumer products to be safe and our obligations to ensure this.

If you have any concerns about our products, you can contact us on ProductSafety@springernature.com

In case Publisher is established outside the EU, the EU authorized representative is:

Springer Nature Customer Service Center GmbH
Europaplatz 3
69115 Heidelberg, Germany

Batch number: 10755329

Printed by Printforce, the Netherlands